MEDICAL
INTELLIGENCE
UNIT

Molecular Mechanisms of Phagocytosis

Carlos Rosales, Ph.D.

Universidad Nacional Autónoma de México
Mexico City, Mexico

LANDES BIOSCIENCE / EUREKAH.COM
GEORGETOWN, TEXAS
U.S.A.

SPRINGER SCIENCE+BUSINESS MEDIA
NEW YORK, NEW YORK
U.S.A.

MOLECULAR MECHANISMS OF PHAGOCYTOSIS

Medical Intelligence Unit

Landes Bioscience / Eurekah.com
Springer Science+Business Media, Inc.

ISBN: 0-387-25419-6   Printed on acid-free paper.

Springer Science+Business Media, Inc., 233 Spring Street, New York, New York 10013, U.S.A.
http://www.springeronline.com

Please address all inquiries to the Publishers:
Landes Bioscience / Eurekah.com, 810 South Church Street, Georgetown, Texas 78626, U.S.A.
Phone: 512/ 863 7762; FAX: 512/ 863 0081
http://www.eurekah.com
http://www.landesbioscience.com

Printed in the United States of America.

9 8 7 6 5 4 3 2 1

Library of Congress Cataloging-in-Publication Data

Molecular mechanisms of phagocytosis / [edited by] Carlos Rosales.
     p. ; cm. -- (Medical intelligence unit)
  Includes bibliographical references and index.
  ISBN 0-387-25419-6
  1. Phagocytosis. I. Rosales, Carlos. II. Title. III. Series: Medical intelligence unit (Unnumbered : 2003)
[DNLM: 1. Phagocytosis--physiology. 2. Molecular Biology. QW 690 M718 2005]
QR187.P4M64 2005
616.07'99--dc22

                              2005005139

For Roberto and Arturo

CONTENTS

Preface xi

1. **Diversity in Phagocytic Signaling: A Story of Greed, Sharing, and Exploitation** 1
   Erick García-García
   Phagocytic Receptors (Greed) 2
   Phagocytic Signaling (Sharing) 6
   Cytoskeleton Dynamics (Exploitation) 12

2. **Phagocytosis and Immunity** 23
   Steven Greenberg
   Diversity of Phagocytic Receptors 23
   Mechanisms of Phagocytic Signaling 24
   Fate of Engulfed Pathogens—Us vs. Them 25
   Phagocytosis and the MHC Class II Pathway 25
   Phagocytosis and the MHC Class I Pathway-Intracellular
      Pathogens and Tumors 26
   Ingestion of Dead Cells and Regulation of Inflammation;
      The Power of "Negative Signaling" 27
   Phagocytosis and Autoimmunity 28

3. **Fc Receptor Phagocytosis** 33
   Randall G. Worth and Alan D. Schreiber
   Fc Receptor Structure 33
   Receptor Activation 34
   Phagocytic Cup Formation 35
   Additional Signal Transduction Molecules: Lipid and Kinases 36
   Calcium Signaling 37
   Phagosome Formation and Fusion 38
   Fc Receptor Interacting Proteins 39
   Inhibitory/Regulatory Fc Receptors 40

4. **Complement Receptors, Adhesion, and Phagocytosis** 49
   Eric Brown
   The Complement Cascade 49
   Complement Receptors 51
   Regulation of Integrin Function 52
   Inside-Out Signaling and Phagocytosis 53
   Complement Receptor Mediated Phagocytosis 54
   Cooperation between IgG Fc Receptors and Complement Receptors ... 55
   Integrin Regulation of Phagocytosis 55

5. **Adding Complexity to Phagocytic Signaling:
   Phagocytosis-Associated Cell Responses and Phagocytic Efficiency** 58
   Erick García-García and Carlos Rosales
   Regulation of Phagocytosis-Associated Cell Responses 59
   Modulation of Phagocytic Efficiency 62

6. **Small GTP Binding Proteins and the Control of Phagocytic Uptake** 72
    Agnès Wiedemann, Jenson Lim and Emmanuelle Caron
    Rho Proteins and Actin Polymerisation 73
    Variations and Anomalies 75
    Ras Proteins and Receptor Activation 76
    Arfs and Rabs and the Delivery of Membrane
        to Forming Phagosomes 78

7. **Regulation of Phagocytosis by FcγRIIb and Phosphatases** 85
    Susheela Tridandapani and Clark L. Anderson
    Signaling Mechanisms Initiated by Macrophage
        ITAM-Associated (Activating) FcγR 85
    Negative Regulation by the ITIM-Bearing FcγRIIb 87
    The SH2 Domain Containing Inositol 5' Phosphatase SHIP-1 89
    The Inositol Phosphatase SHIP-2 90
    PTEN (Phosphatase and Tensin Homologue
        on Chromosome 10) 90
    The Protein Tyrosine Phosphatase SHP-1 91

8. **Phospholipases and Phagocytosis** 97
    Michelle R. Lennartz
    Phagocytosis 97
    Phospholipases 98
    Phospholipase C 99
    Phospholipase D 102
    Lipid Phosphate Phosphohydrolase (LPP) 105
    Phospholipase A2 (PLA2) 105
    cPLA2 and Eicosanoid Production 108
    Sphingomyelinases 108
    Lipid Kinases 110

9. **Calcium Signaling during Phagocytosis** 117
    Alirio J. Melendez
    Receptors Involved in Phagocytosis 118
    A Variety of Nonprofessional Phagocytic Receptors 119
    Calcium the Ubiquitous Second Messenger 120
    What Couples Phagocytic-Receptor Activation to Rise
        on Intracellular Ca^{2+} 122
    Downstream Events Triggered by Ca^{2+} following
        Phagocytic-Receptor Activation 123
    How Might Intracellular Ca^{2+} Control Actin
        Depolymerization during Phagocytosis? 124
    Role of Ca^{2+} on the Maturation of Phagosomes
        during Phagocytosis? 125

10. **Phagosome Maturation** 133
    William S. Trimble and Marc G. Coppolino
    Endocytosis: Snares, Rabs and Tethers 133
    Formation of a Phagosome 136
    Phagosome Maturation 139

Index 151

EDITOR

Carlos Rosales
Universidad Nacional Autónoma de México
Mexico City, Mexico
Chapter 5

CONTRIBUTORS

Clark L. Anderson
Department of Internal Medicine
and the Dorothy M. Davis Heart
and Lung Research Institute
The Ohio State University
Columbus, Ohio, U.S.A.
Chapter 7

Eric Brown
Program in Microbial Pathogenesis
and Host Defense
University of California, San Francisco
San Francisco, California, U.S.A.
Chapter 4

Emmanuelle Caron
Department of Biological Sciences
Centre for Molecular Microbiology
and Infection
Imperial College London
London, United Kingdom
Chapter 6

Marc G. Coppolino
Department of Chemistry
and Biochemistry
University of Guelph
Guelph, Ontario, Canada
Chapter 10

Erick García-García
Immunology Department
Instituto de Investigaciones Biomédicas
Universidad Nacional Autónoma
de México
Mexico City, Mexico
Chapters 1, 5

Steven Greenberg
Columbia University
Departments of Medicine
and Pharmacology
New York, New York, U.S.A.
Chapter 2

Michelle R. Lennartz
Center for Cell Biology and Cancer
Research
Albany Medical College
Albany, New York, U.S.A.
Chapter 8

Jenson Lim
Department of Biological Sciences
Centre for Molecular Microbiology
and Infection
Imperial College London
London, United Kingdom
Chapter 6

Alirio J. Melendez
Department of Physiology
Faculty of Medicine
National University of Singapore
Singapore
Chapter 9

Alan D. Schreiber
Division of Hematology and Oncology
Department of Medicine
University of Pennsylvania School
of Medicine
Philadelphia, Pennsylvania, U.S.A.
Chapter 3

Susheela Tridandapani
Department of Internal Medicine
  and the Dorothy M. Davis Heart
  and Lung Research Institute
The Ohio State University
Columbus, Ohio, U.S.A.
Chapter 7

William S. Trimble
Program in Cell Biology
Hospital for Sick Children
  and Department of Biochemistry
University of Toronto
Toronto, Ontario Canada
Chapter 10

Agnès Wiedemann
Department of Biological Sciences
Centre for Molecular Microbiology
  and Infection
Imperial College London
London, United Kingdom
Chapter 6

Randall G. Worth
Division of Hematology and Oncology
Department of Medicine
University of Pennsylvania School
  of Medicine
Philadelphia, Pennsylvania, U.S.A.
Chapter 3

PREFACE

Phagocytosis, the internalization of large particles by cells, is present in unicellular organisms, which use phagocytosis to eat, all the way to complex pluricellular animals in which special phagocytic cells represent a fundamental part of defense mechanisms. Phagocytosis was first described by the Russian scientist Elie Metchnikoff in the late 1800s, nearly 120 years ago, and it has been studied ever since. In spite of this, only in recent times we began to really understand the molecular basis of this cell function. Because phagocytosis is a very complex process, research about it is done in multiple fronts. So our current knowledge about phagocytosis is found in many different places, mainly articles in scientific journals. This book represents an effort to bring together much of the recent information available about this important cell function. To achieve this, I invited several brilliant scientists who actually investigate phagocytosis on a daily basis, to write chapters. Their contributions resulted in this book, which now puts their knowledge in a single place.

This book describes the various steps of the phagocytic process from initial cell contact, through internalization of the foreign particle, to the final phagosome formation where the phagocytosed particle is destroyed. Each chapter deals with one of these steps and emphasizes the molecules that participate at each step of the process. We begin by providing a general overview of phagocytosis in the chapter "Diversity in phagocytic signaling". Next the chapter "Phagocytosis in immunity" talks about the important role of this cell function in the immune system. The chapters "Fc Receptors and phagocytosis" and "Complement receptors, adhesion, and phagocytosis" describe the major groups of phagocytic receptors, namely receptors for antibodies and complement. Then the chapter "Adding complexity to phagocytic signaling" points out that phagocytosis is not an all-or-nothing response, but rather a function that varies according to the cell and molecules involved. The following chapters deal with specific groups of molecules that have been identified as important regulators of phagocytosis. The chapter "GTPases in phagocytosis" tells us about these molecules and their role in changing the cytoskeleton to bring about the changes in cell shape needed during phagocytosis. The chapter "Role of FcRIIb and phosphatases in phagocytosis" describes the negative effect of this receptor and phosphatases on this process. The chapter "Phospholipases and phagocytosis" indicates how these enzymes participate in activating phagocytic signals and in remodeling the cell membrane. The chapter "Calcium signaling in phagocytosis" describes all that we know today about the role of this ion during the phagocytic process. Finally, the chapter "Phagosome maturation" tells us how phagocytosis continues far beyond the internalization of the particle and describes how the phagosome is formed.

Each chapter can be read independently, but all chapters are connected to bring together several points of view on the same biological function. The book provides in this way a complete modern vision of phagocytosis.

Carlos Rosales
January 2005

CHAPTER 1

Diversity in Phagocytic Signaling:
A Story of Greed, Sharing, and Exploitation

Erick García-García

Abstract

Phagocytosis is the process whereby cells engulf large particles. Phagocytosis is triggered by the interaction of opsonins covering the surface of a phagocytic target with specific phagocyte receptors. In multicellular organisms phagocytosis participates in tissue remodeling and contributes to homeostasis. Higher organisms possess various phagocytic systems. Each system is composed of a series of ligands, specific receptors, and signaling pathways that culminate in particle internalization and destruction. The best studied phagocytic system is that of the receptors that bind to the Fc portion of immunoglobulins. Other phagocytic systems include phagocytosis of complement-opsonized particles, and phagocytosis of apoptotic cells. The signaling pathways elicited by many phagocytic receptors are complex and diverse. Comparison between the signaling pathways elicited by many phagocytic receptors shows that phagocytic signaling pathways share many elements. Shared signaling molecules include tyrosine kinases, lipid kinases, phospholipases, and serine/threonine kinases. Additionally, all phagocytic signaling pathways activate cytoskeleton-remodeling molecules. The dynamic nature of the cytoskeleton is thus exploited by all phagocytic systems to achieve particle internalization. In this review I will discuss the connections between the various signaling pathways of different phagocytic systems, and the regulation of cytoskeleton dynamics as a means to achieve particle internalization.

Introduction

Phagocytosis is the process whereby cells engulf large particles, usually over 0.5 μm in diameter. In multicellular animals phagocytosis participates in homeostasis and tissue remodeling. Phagocytosis plays an essential role in host defense mechanisms through the uptake and destruction of infectious pathogens, and contributes to inflammation and the immune response.[1]

The immune system has a specialized subset of cells, named professional phagocytes, equipped for rapidly and efficiently ingesting invading microorganisms at sites of inflammation. These phagocytes are neutrophils and macrophages.[2] Monocytes (the macrophage precursors) are often included among the professional phagocytes, though they display a lower phagocytic response than neutrophils and macrophages.[1,2] Other cell types, such as dendritic cells, also display phagocytic activity, mainly towards apoptotic cells.[3,4] Outside the immune system, other cell types are also capable of ingesting apoptotic cells. These cell types include fibroblasts,[5] microglia,[6] lens epithelial cells,[7] and other epithelial cell types.[8]

Molecular Mechanisms of Phagocytosis, edited by Carlos Rosales. ©2005 Eurekah.com
and Springer Science+Business Media.

Phagocytosis is triggered by the interaction of opsonins on the surface of the particle to be internalized with specific receptors on the surface of the phagocyte. Most work regarding the regulation of phagocytosis has been done on Fc receptors (FcRs) and complement receptors (CRs). Very recently specific receptors mediating the ingestion of apoptotic cells have been identified, and their signaling pathways are now being characterized. Comparision between the signaling pathways elicited by different phagocytic receptors shows that these signaling pathways share many elements. These include tyrosine kinases, lipid kinases, phospholipases, and serine/threonine kinases.

Regardless of the phagocytic receptor involved, particle internalization requires the exploitation of the dynamic nature of the cytoskeleton. This is achieved through the activation of signaling molecules that activate a series of cytoskeleton-remodeling molecules. Cytoskeleton-remodeling molecules implicated in phagocytosis include GTPases of the Rho family, guanine nucleotide-exchage factors that regulate GTPases, actin nucleation promoting factors, the actin-nucleation complex Arp2/3, and molecular motors of the myosin family. Regulation of cytoskeleton dynamics during phgaocytosis is necessary for pseudopod extension around the particle being internalized, and for myosin-driven phagosome internalization.

Phagocytic Receptors (Greed)

Phagocytes, and specially professional phagocytes, are greedy cells. They have evolved to express a wide array of membrane receptors that allow them to recognize an even wider array of molecular determinants on phagocytic targets. Though greedy, phagocytic cells play by the numbers. Their phagocytic abilities are tightly controlled, so that cells will efficiently respond to activation and maturation signals; and will specifically recognize necrotic or apoptotic cells during tissue remodeling and wound healing. A brief description of known phagocytic receptors follows.

Fc Receptors

Fc receptors (FcRs) recognize the Fc portion of immunoglobulins (Fig. 1A), and are expressed on many cell types of the immune system.[1] Receptors for IgG (FcγR), IgE (FcεR) and IgA (FcαR) have been described.[1] Interaction of FcRs with their immunoglobulin ligands triggers a wide series of leukocyte responses including phagocytosis, respiratory burst, antibody-dependent cell-mediated cytotoxicity, release of pro-inflammatory mediators, and production of cytokines.[1,9] The cellular response initiated by FcR stimulation depends on the particular receptor stimulated, and on the cell type that expresses it.[9] Among FcRs, only FcαR and FcγRs are capable of mediating phagocytosis.[10,11]

FcαR is expressed in neutrophils, monocytes, and macrophages.[10] FcγRs are expressed differentially in many cell types of the immune system. There are three classes of FcγRs: FcγRI, FcγRII, and FcγRIII. Each class consists of several receptor isoforms that are the product of different genes and splicing variants.[11] Class I FcγR is expressed in monocytes, macrophages, and interferon-γ-stimulated neutrophils.[11] The class II FcγR has two members. FcγRIIA is expressed mainly in phagocytes and natural killer cells, whereas FcγRIIB is expressed mainly in T and B lymphocytes.[11] Phagocytes also inducibly express FcγRIIB, but its expression negatively regulates phagocytosis.[12] Class III FcγR is composed of two members.[11] FcγRIIIA is expressed in macrophages, and in monocytes, whereas FcγRIIIB is expressed exclusively in neutrophils. In contrast to FcγRIIIA and all the other FcRs, FcγRIIIB lacks a transmembrane region and a cytoplasmic tail. This receptor is anchored to the membrane by a glycophosphatidylinositol moiety.[11] Though the neutrophil isoform FcγRIIIB is capable of inducing calcium signaling and actin polymerization, its role in phagocytosis is still controversial.[1]

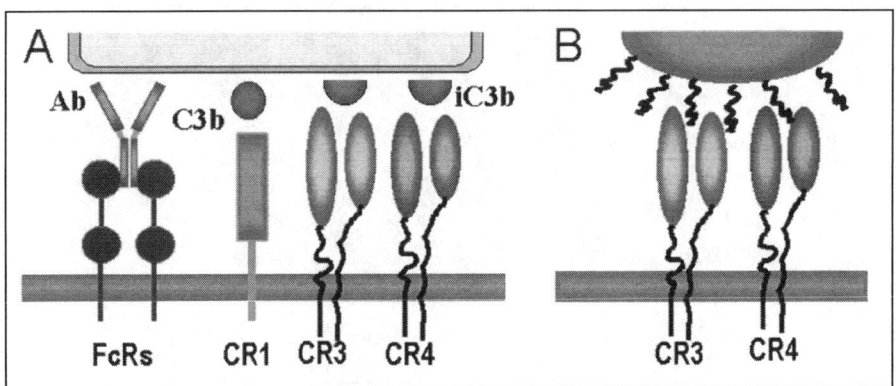

Figure 1. Immunoglobulin and complement receptors. A) Opsonin-dependent phagocytosis is mediated by immunoglobulin (Ig) receptors (FcRs), that bind IgG or IgA; or by complement receptors (CR1, CR3, and CR4), that recognize components of the complement cascade (C3b, or inactive C3b (iC3b)) deposited on the surface of phagocytic targets. B) CR3 and CR4 also mediate nonopsonic phagocytosis of microorganisms through their interaction with sugar ligands (SL).

Complement Receptors

Complement Receptors (CRs) recogninze components of the complement cascade, deposited on the surface of phagocytic targets (Fig. 1A). There are three classes of CRs: CR1, CR3, and CR4. Complement receptor 1 (CR1) is expressed on erythrocytes, phagocytes, and lymphocytes.[1] CR1 recognizes the complement component C3b (Fig. 1A). CR1, however, is unable to trigger phagocytosis of C3b-opsonized particles unless the phagocyte is preactivated by an additional stimulus.[1] CR3 and CR4 are members of the integrin family of receptors.[13] Integrins comprise a large family of membrane receptors consisting of heterodimers of α and β chains. CR3 (αMβ2) and CR4 (αXβ2) are members of the leukocyte-specific β2 integrin subfamily. These receptors are able to recognize the inactive complement component C3b (iC3b) (Fig. 1A). CR4 is expressed poorly in neutrophils and monocytes, and its expression increases upon monocyte-to-macrophage differentiation. CR4 mediates phagocytosis of iC3b-opsonized particles[14] (Fig. 1A), and appears to be important for the nonopsonic, sugar ligand-dependent phagocytosis of various microorganisms[15,16] (Fig. 1B). CR3 is abundantly expressed in monocytes, and its expression is up-regulated upon monocyte-to-macrophage differentiation.[1] Neutrophils express CR3 at low levels, but posses a large intracellular pool of the receptor, whose externalization can be induced upon cell activation.[1,13] CR3 mediates phagocytosis of iC3b-opsonized particles,[1,17] but is also able to recognize intercellular adhesion molecule-1 (ICAM-1), fibrinogen, and coagulation factor X.[17] It is now well accepted that CR3 also possesses a ligand binding site with lectin properties.[18] This lectin site allows CR3 to recognize yeast polysaccharide wall[19] and various microorganism-derived sugar ligands[15,17] (Fig. 1B). The CR3 lectin site is different from the CR3 domain that supports binding of iC3b, ICAM-1, fibrinogen and factor X.[20,21]

Receptors for Apoptotic Cells

Apoptosis is the process of programmed death that cells undergo during development, tissue remodeling, and wound healing. Phagocytic ingestion is the ultimate fate of cells undergoing apoptosis. Ingestion of apoptotic cells occurs very rapidly. This prevents the exposure of surrounding cells and tissues to the potentially harmful contents of the dying cell. Phagocytes specifically ingest apoptotic cells sparing healthy cells. This indicates that apoptotic cells are

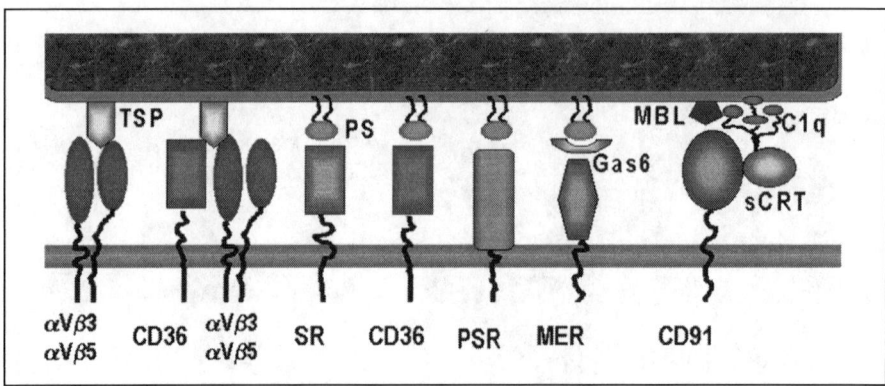

Figure 2. Receptors for apoptotic cells. Integrin receptors αVβ3 and αVβ5 bind thrombospondin (TSP) deposited on the surface of apoptotic cells. CD36 can cooperate with integrins for TSP recognition. Scavenger receptors (SR), CD36, the phosphatidylserine receptor (PSR), and the MER-Gas6 complex recognize phosphatidylserine, exposed on the outer leaflet of the apoptotic cell membrane. CD91 binds a complex composed of surface calreticulin (sCRT), mannose-binding lectin (MBL), and complement component C1q, coating the membrane of apoptotic cells.

somehow targeted for recognition by phagocytes. Several alterations on the surface of apoptotic cells have been described.[22,23] These include alterations on glycosylation patterns, alterations on surface proteins, complement deposition, and loss of phospholipid asymmetry. This last event results in the expression of phosphatidylserine on the outer leaflet of the plasma membrane.[22,23] Surface expression of phosphatidylserine has been demonstrated to be of great importance for apoptotic cell clearance.[24,25] Several receptors for apoptotic cell recognition have been described.[26] These receptors include integrins, scavenger receptors, CD91, the receptor tyrosine kinase MER, the phosphatidylserine receptor, and complement receptors.

Integrins

Various members of the integrin family have implicated in apoptotic cell recognition by macrophages,[27] microglia,[6] dendritic cells,[3,4] and retinal epithelial cells.[28] These include the complement receptors CR3 and CR4,[14] and integrins αVβ3[3,27-31] and αVβ5[4,32] (Fig. 2). In some systems these integrins appear to have a role in the phagocytic process as adhesion molecules, while stimulation of CD36, or of the phosphatidylserine receptor, is the triggering event for apoptotic cell internalization[33,34] (Fig. 3). However, integrin-mediated phagocytosis of apoptotic cells can also occur independently of accessory molecules, by thrombospondin (a molecule secreted by leukocytes at sites of inflammation, that augments cell-to-cell adhesion) recognition on the surface of apoptotic cells[35] (Fig. 2).

Scavenger Receptors

Scavenger receptors are a family of transmembrane receptors that recognize polyanionic ligands, phosphatidylserine, and chemically modified proteins.[23] Several members of the scavenger receptor family mediate apoptotic cell phagocytosis, possibly by direct recognition of phosphatidylserine[23] (Fig. 2). Cell types ingesting apoptotic cells via scavenger receptors include rat Sertolli cells,[36,37] thymic nurse cells,[38] and macrophages.[39] CD36 is a member of the scavenger receptor family with a predominant role in apoptotic cell phagocytosis.[26] CD36 is capable of directly recognizing phosphatidylserine[40,41] (Fig. 2). CD36 also cooperates with the phosphatidylserine receptor,[30] and is required for apoptotic cell ingestion mediated by integrins

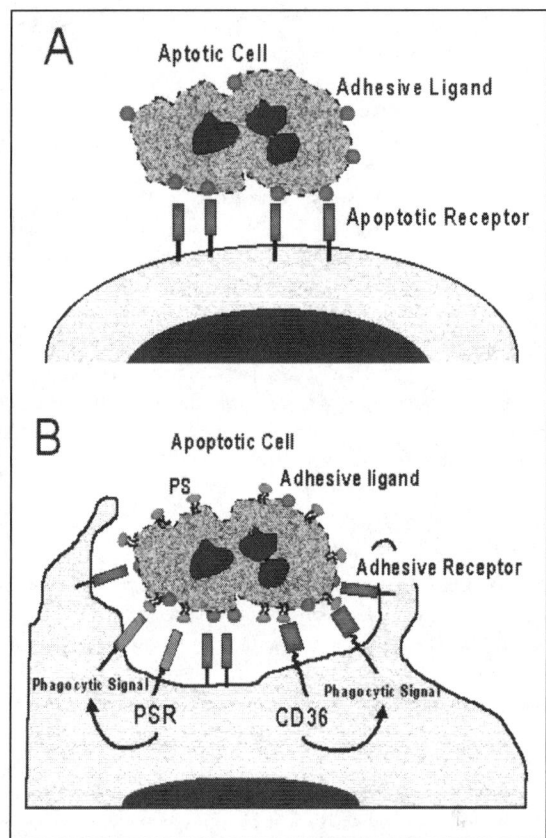

Figure 3. Phosphatidylserine recognition triggers the internalization of apoptotic cells. A) Apoptotic cells can be recognized by phagocytes through the interaction of adhesive ligands on their surface with adhesive receptors on the membrane of phagocytes. This interaction, however, is not sufficient to trigger apoptotic cell internalization. B) Recognition of phosphatidylserine (PS) on the outer leaflet of the apoptotic cell membrane by either the phosphatidylserine receptor (PSR) or CD36 triggers internalization.

αVβ3 and αVβ5[4,29,30] (Fig. 2). Additionally, CD36 mediates phagocytosis through a tripartite complex consisting of CD36, the integrin αVβ3, and thrombospondin[29,42] (Fig. 2).

Phosphatidylserine Receptor

The phosphatidylserine receptor is a transmembrane protein that specifically recognizes phosphatidylserine on the surface of apoptotic cells[5] (Fig. 2). The phosphatidylserine receptor appears to be sufficient to mediate apoptotic cell phagocytosis by professional phagocytes.[5] Additionally, its transfection into a nonphagocytic B-cell line rendered this cells capable of apoptotic cell internalization.[5] The role of the phosphatidylserine receptor as a trigger for apoptotic cell internalization comes from evidence showing that macrophages cannot internalize adhered apoptotic cells, unless the phosphatidylserine receptor is stimulated[34] (Fig. 3). In these cells, however, the phosphatidylserine receptor alone is unable to mediate adhesion.[34] The role of the phosphatydilserine receptor as an internalization-trigger is stressed by the fact that its stimulation induces the ingestion of apoptotic cells adhered to phagocytes through phagocytosis-unrelated molecules such as CD59, or even MHC class I molecules.[34] It is thought that stimulation of the phosphatidylserine receptor induces macropinocytosis, resulting in the internalization of the particles adhered to the phagocyte at the moment of phosphatidylserine receptor stimulation[34] (Fig. 3).

MER

MER is a receptor with tyrosine kinase activity, member of the Axl/MER/Tyro3 receptor family.[43] MER mediates phagocytosis of apoptotic cells by indirectly recognizing phosphatidylserine, through its ligand Gas6[44,45] (Fig. 2). MER-mediated phagocytosis of apoptotic cells has been observed in macrophages,[46] and in retinal pigment epithelial cells.[44,45]

CD91

CD91, also known as low density lipoprotein receptor-related protein, or macroglobulin receptor, is a transmembrane protein that recognizes apoptotic cells through its interaction with calreticulin[47,48] (Fig. 2). Calreticulin is an endoplasmic reticulum-resident molecule that binds complement component C1q, surfactant proteins A and D, and mannose binding lectin, deposited on the surface of apoptotic cells.[47,48]

Phagocytic Signaling (Sharing)

Much work remains to be done in describing the signaling pathways elicited by many phagocytic receptors. In particular a detailed understanding of the signaling pathways regulating CR-mediated phagocytosis, and phagocytosis of apoptotic cells is still lacking. From the available evidence it becomes clear that phagocytosis is regulated by key molecules such as tyrosine kinases, protein kinase C, phospholipases, and phosphatidylinositol 3-kinase; that are shared by most phagocytic systems (Figs. 4-6).

Early Phagocytic Signaling

The signaling pathways elicited by a wide array of membrane receptors involve tyrosine phosphorylation cascades, followed by activation of many down-stream signaling molecules. Phagocytosis involves tyrosine kinases in many cases, although the need for tyrosine kinases may not be a requirement for all phagocytic systems (Figs. 4-6).

Phagocytosis mediated by FcRs is largely mediated by kinases of the Src and Syk/Zap70 families[49] (Fig. 4). FcR ligand interaction induces the phosphorylation of specific tyrosine residues located within special amino acid motifs, named ITAMs (for immunoreceptor tyrosine-based activation motifs),[49] on the cytoplasmic portion of FcγRIIA, and on accessory chains that associate with other FcRs.[10,11] ITAM phosphorylation by enzymes of the Src tyrosine-kinase family[50] promotes Syk docking to ITAMs, and Syk activation[51] (Fig. 4). There is abundant genetic and biochemical evidence supporting the role for Src and Syk family kinases in the regulation of FcγR-mediated phagocytosis.[50,51] Although some members of the Src kinase family have been found associated to specific FcRs[50] it is currently not clear whether this association is related to receptor-specific signaling pathways. It is also possible that all kinases of the Src family converge into a common signaling pathway that regulates phagocytosis. Clarification of this issue will require further investigation.

Participation of tyrosine kinases in CR-mediated phagocytosis is controversial. Early reports demonstrated that phagocytosis of complement-opsonized zymosan (yeast cell wall) and of complement-opsonized erythrocytes is unaffected by tyrosine kinase inhibitors.[52] This ruled out the participation of tyrosine kinases in this type of phagocytosis. This notion is further supported by the fact that macrophages from Syk[-/-] mice show normal levels of CR-mediated phagocytosis.[53] However, β2 integrin stimulation by adhesive ligands, or by artificial integrin cross-linking with antibodies induces various cellular responses in a Src and/or Syk kinase-dependent manner.[13,16] Additionally, it was reported that CR4-mediated phagocytosis

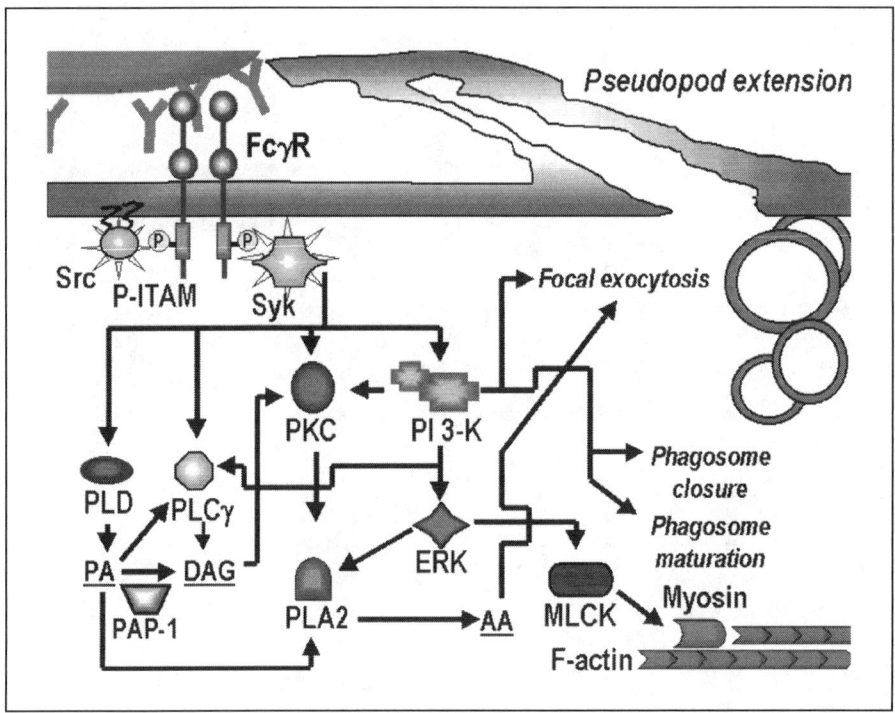

Figure 4. Model for the signaling pathway regulating phagocytosis by Fc receptors. Upon FcR stimulation, Src tyrosine kinases phosphorylate ITAMs on the cytoplasmic portion of FcRs, or on receptor-associated chains. Phosphorylated ITAMs (P-ITAM) promote docking and activation of Syk. After Syk activation several down-stream signaling molecules become activated. Activation of down-stream signaling molecules regulates cytoskeletal rearrangements and membrane remodeling events that are necessary for pseudopod extension, and phagosome closure and internalization (see text for details). PI 3-K: phosphatidylinositol 3-kinase; PKC: protein kinase C; PLCγ: phospholipase Cγ; PLD: phospholipase D; ERK: extracellular-signal regulated kinase; DAG: diacylglycerol; PA: phosphatidic acid; PAP-1: phosphatidic acid-phosphatase 1; AA: arachidonic acid; PLA2: phospholipase A2; MLCK: myosin light chain kinase; F-actin: actin fibers.

of *Mycobacterium tuberculosis* induced tyrosine phosphorylation of several proteins.[54] However, the direct involvement of tyrosine phosphorylation in CR4-mediated phagocytosis was not evaluated.[54] Whether tyrosine kinases are indeed dispensable during CR-mediated phagocytosis remains unclear (Fig. 5).

Phagocytosis of apoptotic cells, on the other hand, appears to depend on tyrosine kinase activity (Fig. 6). It has been reported that tyrosine kinase inhibition has a negative effect on phagocytosis of apoptotic cells by dendritic cells and macrophages.[32,55] Specific inhibition of Src family kinases by PP2 has a similar effect.[55] Scavenger receptor-mediated phagocytosis of apoptotic cells may also involve tyrosine phosphorylation, as it has been shown that the scavenger receptor CD36 interacts with members of the Src kinase family[56,57] (Fig. 6). MER-mediated ingestion of apoptotic cells also involves tyrosine phosphorylation, because the receptor itself has tyrosine kinase activity[43] (Fig. 6). CD91-mediated apoptotic cell phagocytosis also appears to require tyrosine kinases. This is suggested by the fact that interaction of CD91 with its down-stream adaptor GULP depends on a tyrosine-containing NPXYXXL motif located on CD91 cytoplasmic tail.[58] This motif interacts with a phosphotyrosine binding domain on GULP[58] (Fig. 6). Participation of tyrosine kinases in CD91-mediated phagocytosis is further

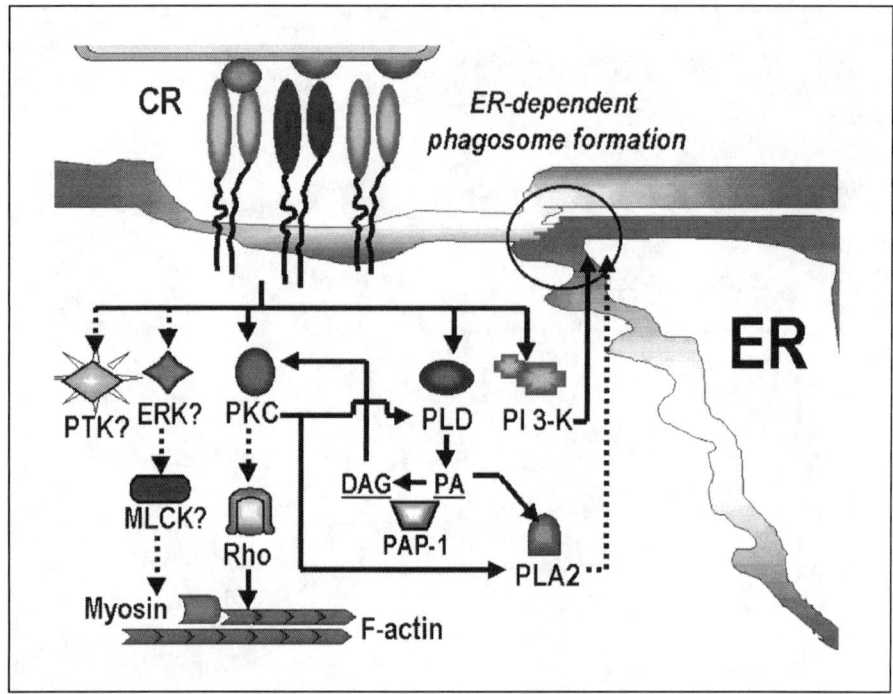

Figure 5. Model for the signaling pathway regulating phagocytosis by complement receptors. Many signaling enzymes have been implicated in the regulation of phagocytosis by complement receptors. Phagocytosis by complement receptors (CR) takes place in the absence of pseudopod extension. Signaling enzymes regulate cytoskeleton rearrangements, and membrane remodeling events (including membrane fusion between the forming phagosome and the endoplasmic reticulum (ER)), necessary for phagosome formation and internalization (see text for details). PTK: protein tyrosine kinases; PI 3-K: phosphatidylinositol 3-kinase; PKC: protein kinase C; PLD: phospholipase D; DAG: diacylglycerol; PA: phosphatidic acid; PAP-1: phosphatidic acid-phosphatase 1; PLA2: phospholipase A2; ERK: extracellular-signal regulated kinase; MLCK: myosin light chain kinase; F-actin: actin fibers. Dotted lines represent possible, yet uncharacterized, connections.

supported by the observation that mutation of the tyrosine within the CD91 NPXYXXL motif suppresses the phagocytic potential of CD91 cytoplasmic domain-containing chimeric receptors.[59] Thus tyrosine kinase activity appears to be a requirement for phagocytosis of apoptotic cells. Despite all these observations, the identification of the particular tyrosine kinases involved in apoptotic cell phagocytosis is still awaiting.

Downstream Phagocytic Signaling

The down-stream signaling events following tyrosine kinase activation have been relatively well characterized in the FcγR phagocytic system.[60] The information regarding down-stream signaling events during CR-mediated phagocytosis, and phagocytosis of apoptotic cells is, in comparison, rather scarce. The next section describes the known roles of various down-stream signaling molecules in the regulation of phagocytosis in the different phagocytic systems.

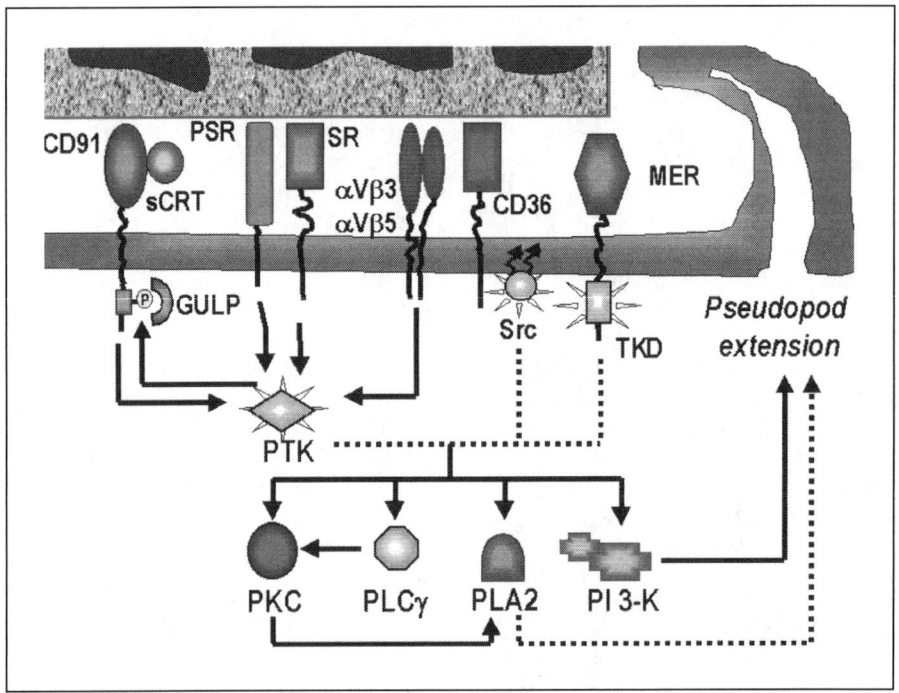

Figure 6. Model for the signaling pathway regulating phagocytosis of apoptotic cells. Phagocytosis of apoptotic cells by several receptors requires protein tyrosine kinase (PTK) activity. CD36 interacts with tyrosine kinases of the Src family, and MER possesses a tyrosine kinase domain (TKD) in its cytoplasmic portion. Activation of tyrosine kinases may be necessary for the activation of down-stream signaling molecules that regulate pseudopod extension (See text for details). SR, scavenger receptors; PSR, phosphatidylserine receptor; sCRT, surface calreticulin; PI 3-K, phosphatidylinositol 3-kinase; PKC, protein kinase C; PLCγ, phospholipase Cγ; DAG, diacylglycerol; PLA2, phospholipase A2. Dotted lines represent possible, yet uncharacterized, connections.

Phosphatidylinositol 3-Kinase

Phosphatidylinositol 3-kinase (PI 3-K) is aknowledged as a key regulator of phagocytic function. It regulates phagocytosis by modulating pseudopod extension, endoplasmic reticulum-dependent phagosome formation, and also by activating many down-stream signaling molecules. All phagocytic systems require PI 3-K activity to accomplish particle internalization (Figs. 4-6).

Pharmacological inhibiton of PI3-K results in phagocytosis arrest during FcγR-mediated phagocytosis.[61-65] The main role for PI 3-K during FcγR-mediated phagocytosis appears to be the regulation of pseudopod extension[63,66] (Fig. 4). This role appears to be fulfilled through the regulation of exocytic membrane insertion at sites of phagocytosis[63] (Fig. 4). Additionally, PI3-K appears to play an important role in the regulation of phagosome closure,[60,67] and in the orchestration of membrane remodeling events that occur during phagosome maturation[66,68] (Fig. 4). PI 3-K may also regulate phagocytosis through indirect activation of extracellular signal-regulated kinase (ERK) (Fig. 4).[64,69] Other possible roles for PI3-K in the regulationof phagocytosis are the activation of some isoforms of protein kinase C (PKC),[70] and the recruitment of phospholipase Cγ (PLCγ), and of the guanine nucleotide-exchange factor Vav [70] to forming phagosomes.

CR-mediated phagocytosis is also reduced by pharmacological inhibition of PI 3-K.[71,72] However, in contrast to FcγR-mediated phagocytosis, CR-mediated phagocytosis occurs in the absence of pseudopod extension. CR3 mediated phagocytosis has been observed to occur acompanied of fusion between the phagosome and the endoplasmic reticulum,[73,74] and this type of phagocytosis requires PI 3-K activity.[73] Thus in the CR phagocytic system PI 3-K may regulate phagosome-endoplasmic reticulum fusion for phagosome formation (Fig. 5).

Phagocytosis of apoptotic cells also depends on PI 3-K activity. Pharmacological inhibition of PI3-K,[55,75] or microinjection of antibodies against the PI3-Kβ catalytic subunit[76] decreased the rate of apoptotic cell ingestion by macrophages. PI3-K may be needed for pseudopod extension during apoptotic cell phagocytosis[75] (Fig. 6).

Protein Kinase C

Enzymes of the PKC family comprise a large family of serine/threonine kinases grouped into four subfamilies, based on structure and cofactor requirements:[77] conventional (α, β_I, β_{II}, γ), novel (δ, ε, η, θ), atypical (ζ,) and the recently described PKCμ, and PKCν.[77] PKC proteins are cytosolic, but may associate to the plasma membrane in response to various activating stimuli.[77] Like PI 3-K, PKC appears to be required by all phagocytic systems to acomplish particle internaliaztion (Figs. 4-6).

Participation of PCK proteins in FcγR-mediated functions has been extensively explored. Pharmacological inhibition or expression of dominant negative isoforms of PKC results in reduced phagocytosis in different cell types.[62,64,78-80] Several PKC isoforms, including PKCβ,[81] PKCγ,[82] PKCδ,[83] and PKCε,[84] are translocated to the membrane after FcγR stimulation. The role of PKC in the regulation of phagocytosis appears to be the activation of down-stream targets, such as ERK[61,64,80,85] and calcium-independent phospholipase A2[86] (Fig. 4).

CR-mediated phagocytosis also requires PKC activity.[52,87] It was shown that neutrophil interaction with zymosan (yeast cell wall) results in membrane translocation of the PKCβII, PKCδ, and PKCξ isoforms.[88] A role for PKC enzymes in phospholipase activation during CR-mediated phagocytosis has also been suggested[89] (Fig. 5). In macrophages PKC-dependent PLA2 activity is observed during phagocytosis of zymosan.[89] It was also reported that PKC activation in neutrophils induced increased phospholipase D (PLD) activity during CR-mediated phagocytosis[87] (Fig. 5). PKC also induces the activation of the GTPase Rho in other systems.[90] Because CR-mediated phagocytosis is Rho dependent,[91] it is also possible that PKC particpates in this type of phagocytosis inducing Rho activation (Fig. 5).

Phagocytosis of apoptotic cells is also reduced by pharmacological inhibition of PKC.[55,92] Phagocytosis of apoptotic cells by macrophages induces membrane translocation of the isoforms PKCβI, PKCβII, PKCδ, PKCε, PKCμ, and PKCξ.[92] PKC appears to be required for apoptotic cell phagocytosis via integrins,[55] or the phosphatidylserine receptor[92] (Fig. 5). The downstream targets of PKC during phagocytosis of apoptotic cells remain to be determined.

Phospholipases

Several phospholipases have been implicated in phagocytosis (Figs. 4-6).

Phospholipase A2

PLA2 mediates arachidonic acid release from phosphatidylcholine, or phosphatidylethanolamine.[93] Arachidonic acid acts as a second messenger during phagocytosis.[94] Although the targets for arachidonic acid have not been identified, this second messenger seems to be important for localized membrane exocytosis and membrane remodeling events[95] that are required for pseudopod extension and phagosome formation.[63]

The participation of PLA2 and arachidonic acid release in FcγR-mediated phagocytosis seems to be the regulation of localized membrane exocytosis necessary for the completion of phagocytosis[63,94-96] (Fig. 4). Similarly, arachidonic acid release has also been observed during CR-mediated phagocytosis.[89] The way arachidonic acid regulates CR-mediated phagocytosis is not known, but it may also regulate membrane remodeling events for phagosome formation (Fig. 5). The participation of PLA2 and arachidonic acid release during apoptotic cell phagocytosis has not been evaluated. However, PLA2 activation has been reported upon scavenger receptor,[97] and αVβ3 integrin[98] stimulation. It is thus likely that PLA2 also participates in apoptotic cell phagocytosis mediated by these receptors (Fig. 6).

Phospholipase Cγ

Phospholipase Cγ (PLCγ) is a phosphoinositide-specific phospholipase that uses phosphatidylinositol-4, 5 bisphosphate for generation of inositol trisphosphate and diacylglycerol.[93] Inositol trisphosphate mediates clacium release from intracellular stores, while diacylglycerol is necessary for activation of several PKC isoforms.[93] However, not all phagocytic systems appear to require PLCγ activity for diacylglycerol formation and PKC activation (Figs. 4-6).

Activation of PLCγ after FcR stimulation has been reported.[99,100] Inhibition of this enzyme results in impaired phagocytosis in macrophages.[101] During FcγR-mediated phagocytosis PLCγ activity may contribute to PKC activation through diacylglycerol production (Fig. 4). In contrast, phagocytosis of complement-opsonized targets appears to occur independently of PLCγ.[87] Diacylglycerol production during CR-mediated phagocytosis, necessary for PKC activation, appears to be achieved indirectly through the activity of phospholipase D[102,103] (Fig. 5). The role for PLCγ in the regulation of apoptotic cell phagocytosis has not been evaluated. However, it has been reported that integrin αVβ3 stimulation induces PLCγ activation.[104] It is thus possible that PLCγ also participates in integrin-mediated phagocytosis of apoptotic cells (Fig. 6).

Phospholipase D

Phospholipase D (PLD) is an enzyme that generates choline and phosphatidic acid from phosphatidylcholine.[93] Phosphatidic acid can be metabolized into diacylglycerol by the enzyme phosphatidic acid phosphatase-1.[93] Thus PLD activation may be an additional pathway for diaylglycerol-dependent PKC activation (Figs. 4,5). Additionally, phosphatidic acid by itself is capable of activating various enzymes, such as PLCγ and PLA2.[93] PLD has been observed to become activated during phagocytosis in different cell types.

Pharmacological inhibition studies show that PLD participates in FcγR-mediated phagocytosis[105] (Fig. 4). In neutrophils, indirect PLD inhibition by ceramide induced a decrease in the rate of FcγR-mediated phagocytosis.[105] PLD may regulate FcγR-mediated phagocytosis by indirect PKC activation, or by phosphatidic acid-dependent PLCγ and PLA2 activation (Fig. 4). Likewise, pharmacological inhibition studies show that PLD is necessary for CR-mediated phagocytosis.[87,102,103,106] PLD has been reported to become activated upon CR stimulation by complement components C3b,[106] iC3b,[87,106] and also by microorganism-derived sugar ligands.[102,103] PLD may regulate CR-mediated phagocytosis by indirect PKC activation, or by phosphatidic acid-dependent PLA2 activation (Fig. 5). The role for PLD in the regulation of apoptotic cells phagoytosis has not been evaluated.

Extracellular Signal-Regulated Kinase

ERK is a serine/threonine kinase involved in signal transduction by a wide variety of receptors including growth factor receptors, integrins, and immune receptors.[9,107] Participation of ERK in phagocytosis has also been reported, although not all phagocytic system seem to require ERK activity (Figs. 4-6).

FcγR-mediated phagocytosis is ERK dependent. Pharmacological inhibition of ERK results in decreased levels of FcγR-mediated phagocytosis in neutrophils, and macrophages.[61,62,78,108] The role of ERK during phagocytosis may be the activation of calcium-dependent PLA2 for arachidonic acid production[60,93] (Fig. 4). Additionally, ERK may also regulate phagocytosis by modulating actin dynamics through myosin activation[60,67] (Fig. 4). Myosins are a large family of ATPases whose interaction with the actin cytoskeleton is thought to provide the mechanical force necessary for phagosome internalization.[60,109] Phosphorylation of some myosin isoforms by myosin light chain kinase (MLCK) results in their activation.[108] Inhibition of MLCK, in neutrophils, results in suppression of phagocytosis.[108] Because in these cells MLCK activation is ERK-dependent, ERK may also regulate particle internalization through MLCK activation, additionally to controlling PLA2 activity (Fig. 4).

Although there is abundant evidence suggesting that integrin signaling occurs via ERK in leukocytes,[16] the role for ERK in CR-mediated phagocytosis has not been directly evaluated (Fig. 6). Because CR-mediated phagocytosis may also require PLA2 activity, and depends on myosin activity, a role for ERK in this type of phagocytosis is likely (Fig. 5).

The role for ERK in phagocytosis of apoptotic cells is less clear. It was reported that ingestion of apoptotic cells by alveolar and peritoneal macrophages induces strong ERK activation.[55] On the other hand, it was also reported that ingestion of apoptotic cells by peritoneal and bone marrow macrophages resulted in decreased resting ERK activity.[110] In this system apoptotic cell ingestion could actually down-regulate cytokine-induced ERK activation.[110] The basis for this contradiction may be related to differences in the apoptosis stage of the cells used for the assays. It has been reported that phagocytosis of late apoptotic cells (as would be encountered during pathological conditions) triggers ERK activation.[111] ERK activity, under this conditions, is necessary for pro-inflammatory cytokine production.[111,112] Clarification of these contradictory observations will require further investigation.

Cytoskeleton Dynamics (Exploitation)

Much work remains to be done in describing the signaling pathways elicited by many phagocytic receptors. However, phagocytosis appears to be regulated by key signaling molecules that are shared by most phagocytic systems (Figs. 4-6). In the end, however, all phagocytic systems must exploit the dynamic nature of the cytoskeleton to achieve particle internalization. Phagocytic signaling pathways, in all systems, are designed to activate many cytoskeleton-remodeling molecules. The actin and microtubule cystoskeletons participate in the phagocytic process in different ways (Figs. 7-9). The actin cytoskeleton is required for pseudopod extension, and as a structural framework for myosin-driven phagosome internalization.[67,113] On the other hand, the microtubule cytoskeleton appears to be necessary for local activation of Rho familiy GTPases at sites of phagosome formation.[114] Different phagocytic systems show particular cytoskeleton requirements to achieve particle internalization.

Cytoskeleton Requirements of Phagocytosis

One of the earliest differences observed between different phagocytic systems were structural. These differences were found through microscopy studies.[75,115] FcγR-mediated phagocytosis and apoptotic cell phagocytosis occur with extensive pseudopod extension surrounding the phagocytic targets[75,115] (Figs. 7,9). CR-mediated phagocytosis, in contrast, occurs in the absence of pseudopod extension, and complement-opsonized phagocytic targets appear to rather sink into the cell[115] (Fig. 8). This differences may reflect the cytoskeletal requirements for particle internalization. The actin cytoskeleton is required for particle internalization by FcγR, and also by receptors for apoptotic cells[75,116] (Figs. 7,9), whereas CR-mediated phagocytosis depends on the actin and microtubule cytoskeletons[52,116] (Fig. 8).

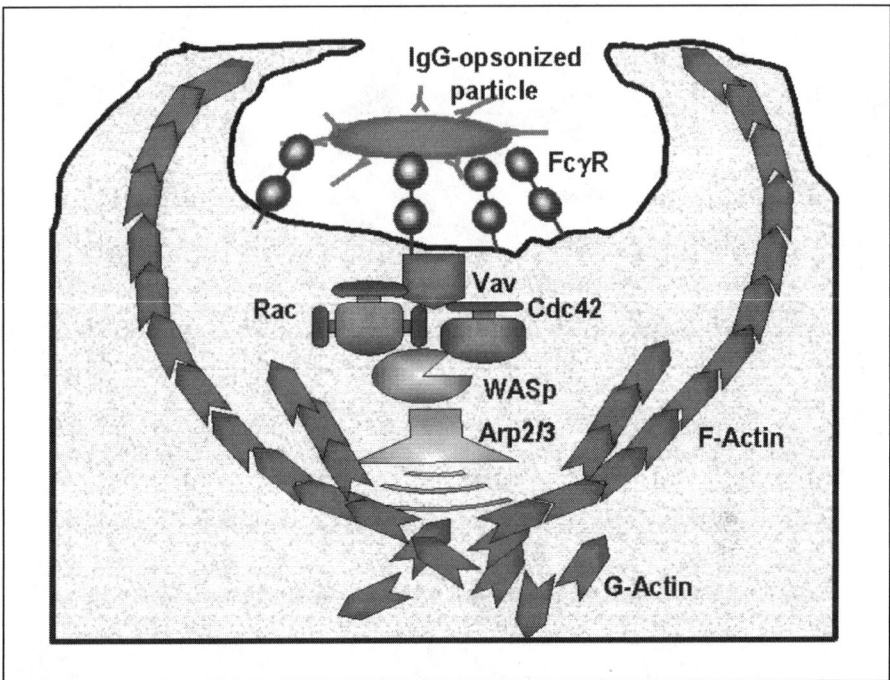

Figure 7. Cytoskeleton regulation during phagocytosis by Fc receptors. Actin polymerization during FcγR-mediated phagocytosis is required for pseudopod extension, and for phagosome internalization. Actin polymerization is regulated by Vav, Rac, and Cdc42. Active Rac/Cdc42 promote the recruitment of the actin nucleation-promoting factor WASp to forming phagosomes. WASp then recruits the actin nucleation complex Arp2/3. G-Actin: globular actin; F-Actin: actin fibers.

Biochemical Regulation of Cytoskeleton Dynamics

Key regulators of cytoskeleton dynamics during phagocytosis include guanine nucleotide-exchange factors (GEFs), adapter molecules with associated GEF activity, Rho family GTPases, actin nucleation promoting factors, the actin nucleation complex Arp2/3, and several myosin isoforms (Figs. 7-9). The role of these molecules in the regulation of phagocytosis is discussed next.

Rho Family GTPases

Rho family GTPases (including Rho, Rac and Cdc42) play a fundamental role in cytoskeleton rearrangements.[67] Rho family GTPases play a key role in the biochemical regulation of the actin,[67,113] and posibly microtubule[117] cytoskeletons during phagocytosis.

The mechanism of activation of Rho family GTPases involves the transition from an inactive, GDP-bound, to an active GTP-bound state.[118] This transition is catalyzed by guanine nucleotide-exchange factors (GEFs).[118] Vav and p190RhoGEF, are GEFs that are implicated in phagocytosis[60,67,119] (Figs. 7-9). Differential use of these GEFs has been observed among phagocytic systems (Figs. 7-9). Vav participates in FcγR-mediated phagocytosis,[120,121] but not in CR-mediated phagocytosis.[120] During FcγR-mediated phagocytosis Vav is recruited to sites of phagosome formation,[121] where it specifically activates Rac[120] (Fig. 7). CR-mediated phagocytosis, on the other hand, appears to depend on p190RhoGEF[119] (Fig. 8). Interestigly,

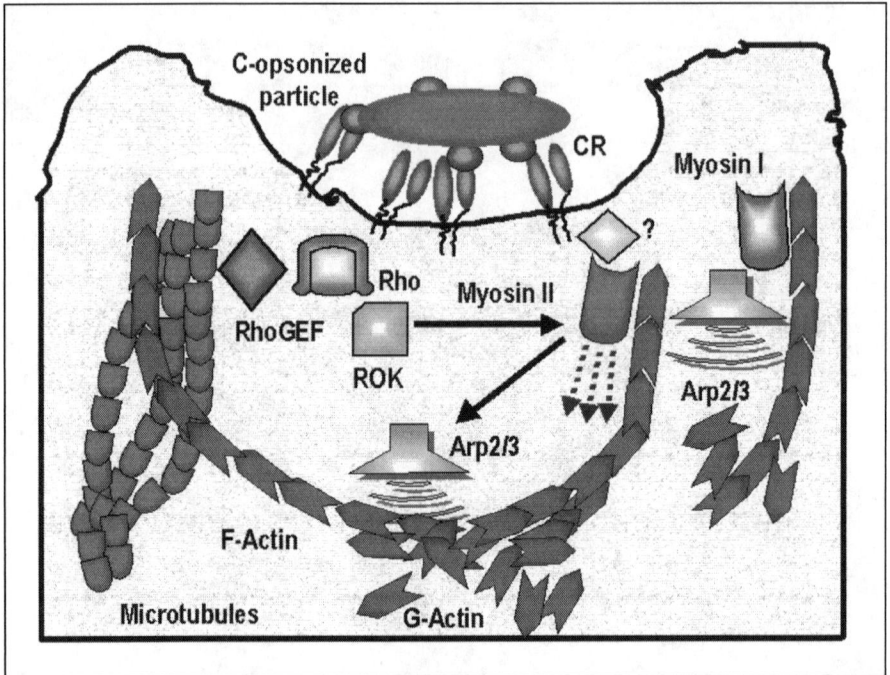

Figure 8. Cytoskeleton regulation during phagocytosis by complement receptors. Phagocytosis by complement receptors requires the actin and microtubule cytoskeletons. The microtubule cytoskeleton may be necessary for activation of Rho at forming phagosomes. Rho activation my be mediated by the Rho-specific activator p190RhoGEF (RhoGEF), that interacts with the microtubule cytoskeleton. Active Rho promotes Arp2/3 recruitment to forming phagosomes, through a signaling pathway involving Rho kinase (ROK), and myosin II. The Arp2/3 complex is also recruited through its interaction with myosin I. Myosins generate the contractile force required for phagosome internalization. The adapter molecules that link myosins with the phagosome membrane are not known (?). G-Actin: globular actin; F-Actin: actin fibers.

p190RhoGEF has been shown to interact with the microtubule cystoskeleton.[114] It is possible that the microtubule cystoskeleton regulates CR-mediated phagocytosis through p190RhoGEF, by regulating the activation of Rho (Fig. 8). In support for this notion, it was found that the microtubule cytoskeleton can control the avidity of β2 integrins in a Rho-dependent manner.[122] Activation of Rho family GTPases during apoptotic cell phagocytosis, on the other hand, appears to be mediated by phosphorylation-dependent adapter molecules with associated GEF activity, such as p130^{cas}, CrkII, Dock180, and ELMO-1[32,123] (Fig. 9). It has been observed that during apoptotic cell phagocytosis these adapter molecules are recruited to sites of phagosome formation where they promote Rac activation[32,123] (Fig. 9).

FcγR-mediated phagocytosis is regulated by the Rho family members Rac and Cdc42, but apparently not by Rho[91] (Fig. 7). In contrast, phagocytosis of complement-opsonized targets depends on Rho (Fig. 8), but is independent of Rac and Cdc42.[91] However, when the lectin site on CR3 is engaged by microorganism-derived sugar ligands, phagocytosis proceeds in a Rac/Cdc42 dependent manner and is accompanied of extensive pseudopod extension, structurally resembling FcγR-mediated phagocytosis.[124] Apoptotic cell phagocytosis also occurs with extensive pseudopod extension, is Rac/Cdc42-dependent (Fig. 9), and appears to be negatively regulated by Rho.[32,75,125]

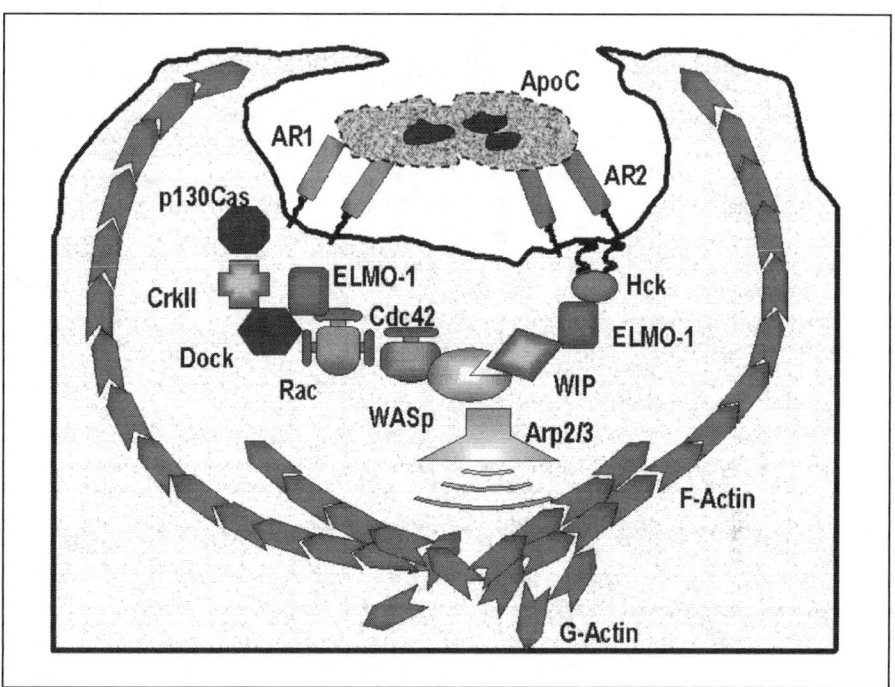

Figure 9. Cytoskeleton regulation during phagocytosis of apoptotic cells. Actin polymerization during phagocytosis of apoptotic cells is necessary for pseudopod extension and phagosome internalization. Two signaling pathways for actin polymerization coexist during phagocytosis of apoptotic cells. It is possible that these signaling pathways are elicited by different sets of receptors (AR1 and AR2). One pathway involves the adapter molecules p130^{cas} (p130Cas), CrkII, ELMO-1, and Dock180 (Dock). Accumulation of this complex at sites of phagocytosis promotes Rac/Cdc42 activation, that is necessary for WASp-Arp2/3 activation and recruitment. A second pathway for actin polymerization involves the adapter molecule ELMO-1, which is recruited to forming phagosomes through its interaction with the Src family kinase member Hck. ELMO-1 interacts with the WASp-interacting molecule (WIP), thus resulting in WASp-Arp2/3 recruitment and activation at forming phagosomes. G-Actin: globular actin; F-Actin: actin fibers.

Actin Polymerization at Sites of Phagosome Formation

Rho family GTPases have a fundamental role in the regulation of the formation of actin fibers that are necessary for pseudopod extension, and for myosin-driven particle internalization[113,126] (Figs. 7-9). Rho family GTPases promote actin fiber formation at sites of phagocytosis by recruiting actin nucleation-promoting factors (NPFs), and the actin-nucleating complex Arp2/3[117,126-129] (Figs. 7-9). While NPFs induce the activation of the Arp2/3 complex, the latter has a direct role in the formation and branching of new actin fibers[117] (Figs. 7-9). Although the recruitment of NPFs at sites of phagosome formation may achieved through different ways among phagocytic systems, all of them require the activity of the Arp2/3 complex for particle internalization (Figs. 7-9). Arp2/3 recruitment to forming phagosomes can be achieved by its interaction with NPFs, with adapter molecules, or with some myosin isoforms (Figs. 7-9).

In the FcγR phagocytic system, active Rac and Cdc42 at forming phagosomes induce the local recruitment of the hematopoietic-specific NPF WASp (for Wiskott-Aldrich Syndrome protein)[121,128] (Fig. 7). WASp recruitment to forming phagosomes may be mediated through its direct interaction with active Rac/Cdc42[117] (Fig. 7). WASp recruitment then results in Arp2/3 activation at these sites.[126] Active Arp2/3 in turn regulates the formation of actin fibers that are necessary for pseudopod extension[60,67] (Fig. 7).

The participation of WASp in CR-mediated phagocytosis has not been evaluated. However, CR-mediated phagocytois does depend on Arp2/3 recruitment to forming phagosomes[127] (Fig. 8). During CR-mediated phagocytosis the Arp2/3 complex is recruited in a Rho-dependent manner[127] (Fig. 8). In this system Arp2/3 recruitment to forming phagosomes is a down-stream event, occurring after activation of the Rho kinase-myosin II signaling pathway[127] (Fig. 8). Additionally, Arp2/3 may be recruited to forming phagosomes by its interaction with myosin I, the only myosin isoform known to directly interact with Arp2/3[117] (Fig. 8). In support of this notion it was reported that myosin I is found around phagosomes during CR-mediated internalization of zymosan.[130] Because phagocytosis by CRs proceeds in the absence of pseudopod extension, the most likely role for actin fibers in this system, is to provide the structural framework for myosin-driven phagosome internalization (Fig. 8).

Phagocytosis of apoptotic cells requires recruitment of WASp to sites of phagocytosis[129] (Fig. 9). During apoptotic cell phagocytosis WASp recruitment appears to be regulated by ELMO-1,[131] and may be additionally regulated by Rac and Cdc42[75,129] (Fig. 9). In macrophages it was shown that during apoptotic cell phagocytosis ELMO-1 interacts with WASp through a molecular complex consisting of the Src family member Hck, ELMO-1, and the WASp binding partner WIP (for WASp-interacting protein)[131] (Fig. 9). Because apoptotic cell phagocytosis is Rac/Cdc42 dependent,[75] it is also possible that active Rac/Cdc42 at forming phagosomes serves to recruit WASp in a manner analogous to the FcγR phagocytic system (Fig. 9). WASp located at forming phagosomes then recruits the Arp2/3 complex for actin fiber formation, and pseudopod extension (Fig. 9).

Actin Driven Phagosome Internalization

In addition to the role of actin polymerization for pseudopod extension, actin fibers provide the structural framework for the myosin-driven contractile activity, necessary for particle internalization.[130,132,133] This actin-dependent contractile activity is mediated by several myosin isoforms.[109] Myosins are motor-proteins that couple their ATPase activity to movement along actin fibers.[109] By coupling its movement to adapter molecules, myosins can thus regulate the transport of vesicles, organelles and other particles along actin fibers[109] (Fig. 8). Several myosin isoforms have been found located around phagosomes, suggesting that some of them are required for phagosome internalization.[67]

Myosin II appears to be required for FcγR-mediated phagocytosis.[108,132] Pharmacological inhibition of myosin II, or of its upstream activator MLCK results in phagocytosis arrest.[108,132] Other myosin isoforms, including myosin IC, myosin V and myosin IXb, have been found located around phagosomes during FcγR-mediated phagocytosis.[132] It is thus possible that at least one of them participates in phagosome internalization. CR-mediated phagocytosis also depends on myosin II activity[127] (Fig. 8). Myosin I has also been found located around phagosomes during CR-mediated phagocytosis[130] (Fig. 8). Additionally to the possible role for these myosin isofoms in phagosome internalization, they may also participate in actin fiber formation by recruiting the Arp2/3 complex at forming phagosomes[117,127] (Fig. 8). The myosin isoforms required for phagosome internalization during apoptotic cell phagocytosis have not yet been identified.

Conclusion

Phagocytosis is a fundamental cellular function in multicellular organisms. It plays a crucial role in host defense mechanisms, wound healing, and tissue remodeling. Higher organisms posses specialized subsets of cells, named professional phagocytes, that are equipped with a wide series of phagocytic receptors that make them hungry and greedy. This battery of receptors enable phagocytic cells to recognize opsonins, and other molecular determinants on invading microorganisms or apoptotic cells.

The different phagocytic systems share important members of their signaling pathways. These shared signaling elements include tyrosine kinases, lipid kinases, phospholipases, serine/threonine kinases, and actin-binding proteins.

Phagocytic signaling pathways are designed to exploit the dynamic nature of the cytoskeleton as a means to achieve particle internalization. Rho family GTPases play a fundamental role in the regulation of cytoskeleton dynamics, by recruiting and activating many cytoskeleton-regulating molecules at sites of phagocytosis. Cytoskeleton remodeling contributes to pseudopod extension, and to myosin-driven phagosome internalization.

Much work remains to be done in describing the signaling pathways elicited by many phagocytic receptors, and the way phagocytosis is modulated depending on activation or differentiation signals. However, future research in these areas this will be an exciting and rewarding path to walk.

Acknowledgements

This work was supported by grant 36407-M from Consejo Nacional de Ciencia y Tecnología (CONACyT), México. I thank Dr. Carlos Rosales Ledezma fos his critical review of this manuscript.

References

1. Jones SL, Lindberg FP, Brown EJ. Phagocytosis. In: Paul WE, ed. Fundamental Immunology. 4th ed. Philadelphia: Lippincott-Raven Publishers, 1999:997-1020.
2. Rabinovitch M. Professional and nonprofessional phagocytes: An introduction. Trends in Cell Biology 1995; 5:85-87.
3. Rubartelli A, Poggi A, Zocchi MR. The selective engulfment of apoptotic bodies by dendritic cells is mediated by the alpha(v)beta3 integrin and requires intracellular and extracellular calcium. Eur J Immunol 1997; 27:1893-900.
4. Albert ML, Pearce SF, Francisco LM et al. Immature dendritic cells phagocytose apoptotic cells via alphavbeta5 and CD36, and cross-present antigens to cytotoxic T lymphocytes. J Exp Med 1998; 188:1359-68.
5. Fadok VA, Bratton DL, Rose DM et al. A receptor for phosphatidylserine-specific clearance of apoptotic cells. Nature 2000; 405:85-90.
6. Witting A, Muller P, Herrmann A et al. Phagocytic clearance of apoptotic neurons by Microglia/Brain macrophages in vitro: Involvement of lectin-, integrin-, and phosphatidylserine-mediated recognition. J Neurochem 2000; 75:1060-70.
7. Ryeom SW, Sparrow JR, Silverstein RL. CD36 participates in the phagocytosis of rod outer segments by retinal pigment epithelium. J Cell Sci 1996; 109(Pt 2):387-95.
8. Parnaik R, Raff MC, Scholes J. Differences between the clearance of apoptotic cells by professional and nonprofessional phagocytes. Curr Biol 2000; 10:857-60.
9. Sánchez-Mejorada G, Rosales C. Signal transduction by immunoglobulin Fc receptors. J Leukoc Biol 1998; 63:521-533.
10. Monteiro R, Van De Winkel J. IgA Fc receptors. Annu Rev Immunol 2003; 21:177-204.
11. Ravetch JV, Bolland S. IgG Fc receptors. Annu Rev Immunol 2001; 19:275-290.
12. Tridandapani S, Siefker K, Teillaud JL et al. Regulated expression and inhibitory function of FcgRIIb in human monocytic cells. J Biol Chem 2002; 277:5082-5089.
13. Dib K, Andersson T. BETA 2 integrin signaling in leukocytes. Font Biosci 2000; 5:438-51.
14. Takizawa F, Tsuji S, Nagasawa S. Enhancement of macrophage phagocytosis upon iC3b deposition on apoptotic cells. FEBS Lett 1996; 397:269-72.
15. Ofek I, Goldhar J, Keisari Y et al. Nonopsonic phagocytosis of microorganisms. Annu Rev Microbiol 1995; 49:239-76.

16. Lowell CA, Berton G. Integrin signal transduction in myeloid leukocytes. J Leukoc Biol 1999; 65:313-20.

17. Ehlers MR. CR3: A general purpose adhesion-recognition receptor essential for innate immunity. Microbes Infect 2000; 2:289-94.

18. Ross GD, Vetvicka V. CR3 (CD11b, CD18): A phagocyte and NK cell membrane receptor with multiple ligand specificities and functions. Clin Exp Immunol 1993; 92:181-4.

19. Ross GD, Cain JA, Myones BL et al. Specificity of membrane complement receptor type three (CR3) for beta-glucans. Complement 1987; 4:61-74.

20. Thornton BP, Vetvicka V, Pitman M et al. Analysis of the sugar specificity and molecular location of the beta-glucan-binding lectin site of complement receptor type 3 (CD11b/CD18). J Immunol 1996; 156:1235-46.

21. Diamond MS, Garcia-Aguilar J, Bickford JK et al. The I domain is a major recognition site on the leukocyte integrin Mac-1 (CD11b/CD18) for four distinct adhesion ligands. J Cell Biol 1993; 120:1031-43.

22. Elward K, Gasque P. "Eat me" and "don't eat me" signals govern the innate immune response and tissue repair in the CNS: Emphasis on the critical role of the complement system. Mol Immunol 2003; 40:85-94.

23. Platt N, da Silva RP, Gordon S. Recognizing death: The phagocytosis of apoptotic cells. Trends Cell Biol 1998; 8:365-72.

24. Kagan VE, Borisenko GG, Serinkan BF et al. Appetizing rancidity of apoptotic cells for macrophages: Oxidation, externalization, and recognition of phosphatidylserine. Am J Physiol Lung Cell Mol Physiol 2003; 285:L1-17.

25. Fadok VA, Bratton DL, Frasch SC et al. The role of phosphatidylserine in recognition of apoptotic cells by phagocytes. Cell Death Differ 1998; 5:551-62.

26. Krieser RJ, White K. Engulfment mechanism of apoptotic cells. Curr Opin Cell Biol 2002; 14:734-8.

27. Fadok VA, Savill JS, Haslett C et al. Different populations of macrophages use either the vitronectin receptor or the phosphatidylserine receptor to recognize and remove apoptotic cells. J Immunol 1992; 149:4029-35.

28. Hisatomi T, Sakamoto T, Sonoda KH et al. Clearance of apoptotic photoreceptors: Elimination of apoptotic debris into the subretinal space and macrophage-mediated phagocytosis via phosphatidylserine receptor and integrin alphavbeta3. Am J Pathol 2003; 162:1869-79.

29. Stern M, Savill J, Haslett C. Human monocyte-derived macrophage phagocytosis of senescent eosinophils undergoing apoptosis. Mediation by alpha v beta 3/CD36/thrombospondin recognition mechanism and lack of phlogistic response. Am J Pathol 1996; 149:911-21.

30. Fadok VA, Warner ML, Bratton DL et al. CD36 is required for phagocytosis of apoptotic cells by human macrophages that use either a phosphatidylserine receptor or the vitronectin receptor (alpha v beta 3). J Immunol 1998; 161:6250-7.

31. Moodley Y, Rigby P, Bundell C et al. Macrophage recognition and phagocytosis of apoptotic fibroblasts is critically dependent on fibroblast-derived thrombospondin 1 and CD36. Am J Pathol 2003; 162:771-9.

32. Albert ML, Kim JI, Birge RB. Alphavbeta5 integrin recruits the CrkII-Dock180-rac1 complex for phagocytosis of apoptotic cells. Nat Cell Biol 2000; 2:899-905.

33. Finnemann SC, Silverstein RL. Differential roles of CD36 and alphavbeta5 integrin in photoreceptor phagocytosis by the retinal pigment epithelium. J Exp Med 2001; 194:1289-98.

34. Hoffmann PR, deCathelineau AM, Ogden CA et al. Phosphatidylserine (PS) induces PS receptor-mediated macropinocytosis and promotes clearance of apoptotic cells. J Cell Biol 2001; 155:649-59.

35. Hughes J, Liu Y, Van Damme J et al. Human glomerular mesangial cell phagocytosis of apoptotic neutrophils: Mediation by a novel CD36-independent vitronectin receptor/thrombospondin recognition mechanism that is uncoupled from chemokine secretion. J Immunol 1997; 158:4389-97.

36. Shiratsuchi A, Kawasaki Y, Ikemoto M et al. Role of class B scavenger receptor type I in phagocytosis of apoptotic rat spermatogenic cells by Sertoli cells. J Biol Chem 1999; 274:5901-8.

37. Kawasaki Y, Nakagawa A, Nagaosa K et al. Phosphatidylserine binding of class B scavenger receptor type I, a phagocytosis receptor of testicular sertoli cells. J Biol Chem 2002; 277:27559-66.

38. Imachi H, Murao K, Hiramine C et al. Human scavenger receptor B1 is involved in recognition of apoptotic thymocytes by thymic nurse cells. Lab Invest 2000; 80:263-70.
39. Platt N, da Silva RP, Gordon S. Class A scavenger receptors and the phagocytosis of apoptotic cells. Immunol Lett 1999; 65:15-9.
40. Tait JF, Smith C. Phosphatidylserine receptors: Role of CD36 in binding of anionic phospholipid vesicles to monocytic cells. J Biol Chem 1999; 274:3048-54.
41. Ryeom SW, Silverstein RL, Scotto A et al. Binding of anionic phospholipids to retinal pigment epithelium may be mediated by the scavenger receptor CD36. J Biol Chem 1996; 271:20536-9.
42. Savill J, Hogg N, Ren Y et al. Thrombospondin cooperates with CD36 and the vitronectin receptor in macrophage recognition of neutrophils undergoing apoptosis. J Clin Invest 1992; 90:1513-22.
43. Lemke G, Lu Q. Macrophage regulation by Tyro 3 family receptors. Curr Opin Immunol 2003; 15:31-6.
44. Hall MO, Prieto AL, Obin MS et al. Outer segment phagocytosis by cultured retinal pigment epithelial cells requires Gas6. Exp Eye Res 2001; 73:509-20.
45. Hall MO, Obin MS, Prieto AL et al. Gas6 binding to photoreceptor outer segments requires gamma-carboxyglutamic acid (Gla) and Ca(2+) and is required for OS phagocytosis by RPE cells in vitro. Exp Eye Res 2002; 75:391-400.
46. Scott RS, McMahon EJ, Pop SM et al. Phagocytosis and clearance of apoptotic cells is mediated by MER. Nature 2001; 411:207-11.
47. Ogden CA, deCathelineau A, Hoffmann PR et al. C1q and mannose binding lectin engagement of cell surface calreticulin and CD91 initiates macropinocytosis and uptake of apoptotic cells. J Exp Med 2001; 194:781-95.
48. Vandivier RW, Ogden CA, Fadok VA et al. Role of surfactant proteins A, D, and C1q in the clearance of apoptotic cells in vivo and in vitro: Calreticulin and CD91 as a common collectin receptor complex. J Immunol 2002; 169:3978-86.
49. Strzelecka A, Kwiatkowska K, Sobota A. Tyrosine phosphorylation and Fcg receptor-mediated phagocytosis. FEBS Letters 1997; 400:11-14.
50. Korade-Mirnics Z, Corey SJ. Src kinase-mediated signaling in leukocytes. J Leukoc Biol 2000; 68:603-613.
51. Turner MES, Colucci F, Di Santo JP et al. Tyrosine kinase SYK: Essential functions for immunoreceptor signalling. Immunol Today 2000; 21:148-154.
52. Allen LA, Aderem A. Molecular definition of distinct cytoskeletal structures involved in complement- and Fc receptor-mediated phagocytosis in macrophages. J Exp Med 1996; 184:627-37.
53. Kiefer F, Brumell J, Al Alawi N et al. The Syk protein tyrosine kinase is essential for Fcgamma receptor signaling in macrophages and neutrophils. Mol Cell Biol 1998; 18:4209-20.
54. Zaffran Y, Zhang L, Ellner JJ. Role of CR4 in Mycobacterium tuberculosis-human macrophages binding and signal transduction in the absence of serum. Infect Immun 1998; 66:4541-4.
55. Hu B, Punturieri A, Todt J et al. Recognition and phagocytosis of apoptotic T cells by resident murine tissue macrophages require multiple signal transduction events. J Leukoc Biol 2002; 71:881-9.
56. Moore KJ, El Khoury J, Medeiros LA et al. A CD36-initiated signaling cascade mediates inflammatory effects of beta-amyloid. J Biol Chem 2002; 277:47373-9.
57. Huang MM, Bolen JB, Barnwell JW et al. Membrane glycoprotein IV (CD36) is physically associated with the Fyn, Lyn, and Yes protein-tyrosine kinases in human platelets. Proc Natl Acad Sci USA 1991; 88:7844-8.
58. Su HP, Nakada-Tsukui K, Tosello-Trampont AC et al. Interaction of CED-6/GULP, an adapter protein involved in engulfment of apoptotic cells with CED-1 and CD91/low density lipoprotein receptor related protein (LRP). J Biol Chem 2002; 277:11772-9.
59. Patel M, Morrow J, Maxfield F et al. The cytoplasmic domain of LDL receptor-related protein, but not that of the LDL receptor, triggers phagocytosis. J Biol Chem 2003, (Epub ahead of print Aug. 26 2003).
60. Garcia-Garcia E, Rosales C. Signal transduction during Fc receptor-mediated phagocytosis. J Leukoc Biol 2002; 72:1092-108.
61. Raeder EM, Mansfield PJ, Hinkovska-Galcheva V et al. Syk activation initiates downstream signaling events during human polymorphonuclear leukocyte phagocytosis. J Immunol 1999; 163:6785-6793.

62. Yamamori T, Inanami O, Nagahata H et al. Roles of p38MAPK, PKC and PI3-K in the signaling pathways of NADPH oxidase activation and phagocytosis in bovine polymorphonuclear leukocytes. FEBS Lett 2000; 467:253-258.

63. Cox D, Tseng CC, Bjekic G et al. A requirement for phosphatidylinositol 3-kinase in pseudopod extension. J Biol Chem 1999; 274:1240-1247.

64. Garcia-Garcia E, Rosales R, Rosales C. Phosphatidylinositol 3-kinase and extracellular signal-regulated kinase are recruited for Fc receptor-mediated phagocytosis during monocyte to macrophage differentiation. Journal of Leukocyte Biology 2002; 72:107-114.

65. Ninoyima N, Hazeki K, Fukui Y et al. Involvement of phosphatidylinositol 3-kinase in Fcg receptor signaling. J Biol Chem 1994; 269:22732-22737.

66. Booth JW, Trimble WS, Grinstein S. Membrane dynamics in phagocytosis. Semin Immunol 2001; 13:357-364.

67. May RC, Machesky LM. Phagocytosis and the actin cytoskeleton. J Cell Sci 2001; 114:1061-1077.

68. Gillooly DJ, Simonsen A, Stenmark H. Phosphoinositides and phagocytosis. J Cell Biol 2001; 155:15-7.

69. Coxon PY, Rane MJ, Powell DW et al. Differential mitogen-activated protein kinase stimulation by Fcg receptor IIa and Fcg receptor IIIb determines the activation phenotype of human neutrophils. J Immunol 2000; 164:6530-6537.

70. Wymann MP, Sozzani S, Altruda F et al. Lipids on the move: Phosphoinositide 3-kinases in leukocyte function. Immunol Today 2000; 21:260-264.

71. Cox D, Dale BM, Kashiwada M et al. A regulatory role for Src homology 2 domain-containing inositol 5'-phosphatase (SHIP) in phagocytosis mediated by Fcg receptors and complement receptor 3 (aMb2; CD11b/CD18). J Exp Med 2001; 193:61-72.

72. Lutz MA, Correll PH. Activation of CR3-mediated phagocytosis by MSP requires the RON receptor, tyrosine kinase activity, phosphatidylinositol 3-kinase, and protein kinase C zeta. J Leukoc Biol 2003; 73:802-14.

73. Gagnon E, Duclos S, Rondeau C et al. Endoplasmic reticulum-mediated phagocytosis is a mechanism of entry into macrophages. Cell 2002; 110:119-31.

74. Houde M, Bertholet S, Gagnon E et al. Phagosomes are competent organelles for antigen cross-presentation. Nature 2003; 425:402-6.

75. Leverrier Y, Ridley AJ. Requirement for Rho GTPases and PI 3-kinases during apoptotic cell phagocytosis by macrophages. Curr Biol 2001; 11:195-9.

76. Leverrier Y, Okkenhaug K, Sawyer C et al. Class I phosphoinositide 3-kinase p110beta is required for apoptotic cell and Fcgamma receptor-mediated phagocytosis by macrophages. J Biol Chem 2003; 278:38437-42.

77. Dempsey EC, Newton AC, Mochly-Rosen D et al. Protein kinase C isozymes and the regulation of diverse cell responses. Am J Physiol Lung Cell Mol Physiol 2000; 279:L429-L438.

78. Raeder EM, Mansfield PJ, Hinkovska-Galcheva V et al. Sphingosine blocks human polymorphonuclear leukocyte phagocytosis through inhibition of mitogen-activated protein kinase activation. Blood 1999; 93:686-693.

79. Karimi K, Gemmill TR, Lennartz MR. Protein kinase C and a calcium-independent phospholipase are required for IgG-mediated phagocytosis by Mono-Mac-6 cells. J Leukoc Biol 1999; 65:854-862.

80. Breton A, Descoteaux A. Protein kinase C-a participates in FcgR-mediated phagocytosis in macrophages. Biochem Biophys Res Com 2000; 276:472-476.

81. Dekker LV, Leitges M, Altschuler G et al. Protein kinase C-b contributes to NADPH oxidase activation in neutrophils. Biochem J 2000; 347:285-289.

82. Melendez AJ, Harnett MM, Allen JM. FcgRI activation of phospholipase Cg1 and protein kinase C in dibutyryl cAMP-differentiated U937 cells is dependent solely on the tyrosine-kinase activated form of phosphatidylinositol 3-kinase. Immunol 1999; 98:1-8.

83. Brumell JH, Howard JC, Craig K et al. Expression of the protein kinase C substrate pleckstrin in macrophages: Association with phagosomal membranes. J Immunol 1999; 163:3388-3395.

84. Larsen EC, DiGennaro JA, Saito N et al. Differential requirement for classic and novel PKC isoforms in respiratory burst and phagocytosis in RAW 264.7 cells. J Immunol 2000; 165:2809-2817.

85. Karimi K, Lennartz MR. Mitogen-activated protein kinase is activated during IgG-mediated phagocytosis, but it is not required for target ingestion. Inflammation 1998; 22:67-82.

86. Lennartz MR, Yuen AFC, McKenzie Masi S et al. Phospholipase A2 inhibition results in seques-tration of plasma membrane into electronlucent vesicles during IgG-mediated phagocytosis. J Cell Sci 1997; 110:2041-2052.
87. Fallman M, Gullberg M, Hellberg C et al. Complement receptor-mediated phagocytosis is associ-ated with accumulation of phosphatidylcholine-derived diglyceride in human neutrophils. Involve-ment of phospholipase D and direct evidence for a positive feedback signal of protein kinase. J Biol Chem 1992; 267:2656-63.
88. Sergeant S, McPhail LC. Opsonized zymosan stimulates the redistribution of protein kinase C isoforms in human neutrophils. J Immunol 1997; 159:2877-85.
89. Akiba S, Mizunaga S, Kume K et al. Involvement of group VI Ca^{2+}-independent phospholipase A2 in protein kinase C-dependent arachidonic acid liberation in zymosan-stimulated macrophage-like P388D$_1$ cells. J Biol Chem 1999; 274:19906-19912.
90. Mehta D, Rahman A, Malik AB. Protein kinase C-alpha signals rho-guanine nucleotide dissocia-tion inhibitor phosphorylation and rho activation and regulates the endothelial cell barrier func-tion. J Biol Chem 2001; 276:22614-20.
91. Caron E, Hall A. Identification of two distinct mechanisms of phagocytosis controlled by different Rho GTPases. Science 1998; 282:1717-21.
92. Todt JC, Hu B, Punturieri A et al. Activation of protein kinase C beta II by the stereo-specific phosphatidylserine receptor is required for phagocytosis of apoptotic thymocytes by resident mu-rine tissue macrophages. J Biol Chem 2002; 277:35906-14.
93. Lennartz MR. Phospholipases and phagocytosis: The role of phospholipid-derived second messen-gers in phagocytosis. Int J Biochem Cell Biol 1999; 31:415-430.
94. Lennartz MR, Brown EJ. Arachidonic acid is essential for IgG Fc receptor-mediated phagocytosis by human monocytes. J Immunol 1991; 147:621-626.
95. Lennartz MR, Lefkowith JB, Bromley FA et al. Immunoglobulin G-mediated phagocytosis acti-vates a calcium-independent, phosphatidylethanolamine-specific phospholipase. J Leukoc Biol 1993; 54:389-398.
96. Karimi K, Lennartz MR. Protein kinase C activation precedes arachidonic acid release during IgG-mediated phagocytosis. J Immunol 1995; 155:5786-5794.
97. Pollaud-Cherion C, Vandaele J, Quartulli F et al. Involvement of calcium and arachidonate me-tabolism in acetylated-low-density-lipoprotein-stimulated tumor-necrosis-factor-alpha production by rat peritoneal macrophages. Eur J Biochem 1998; 253:345-53.
98. Bhattacharya S, Patel R, Sen N et al. Dual signaling by the alpha(v)beta(3)-integrin activates cyto-solic PLA(2) in bovine pulmonary artery endothelial cells. Am J Physiol Lung Cell Mol Physiol 2001; 280:L1049-56.
99. Azzoni L, Kamoun M, Salcedo TW et al. Stimulation of FcgRIIIA results in phospholipase C-g1 tyrosine phosphorylation and p56^{lck} activation. J Exp Med 1992; 176:1745-1750.
100. Shen Z, Lin C-T, Unkeless JC. Correlation among tyrosine phosphorylation of Sch, p72^{syk}, PLC-g1, and [Ca^{2+}]i flux in FcgRIIA signaling. J Immunol 1994; 152:3017-3023.
101. Botelho RJ, Teruel M, Dierckman R et al. Localized biphasic changes in phosphatidylinositol-4,5-biphosphate at sites of phagocytosis. J Cell Biol 2000; 151:1353-1368.
102. Serrander L, Fallman M, Stendahl O. Activation of phospholipase D is an early event in integrin-mediated signalling leading to phagocytosis in human neutrophils. Inflammation 1996; 20:439-50.
103. Kusner DJ, Hall CF, Schlesinger LS. Activation of phospholipase D is tightly coupled to the phagocytosis of Mycobacterium tuberculosis or opsonized zymosan by human macrophages. J Exp Med 1996; 184:585-95.
104. Nakamura I, Lipfert L, Rodan GA et al. Convergence of alpha(v)beta(3) integrin- and macrophage colony stimulating factor-mediated signals on phospholipase Cgamma in prefusion osteoclasts. J Cell Biol 2001; 152:361-73.
105. Suchard SJ, Mansfield PJ, Boxer LA et al. Mitogen-activated protein kinase action during IgG-dependent phagocytosis in human neutrophils. Inhibition by Ceramide. J Immunol 1997; 158:4961-4967.
106. Fallman M, Andersson R, Andersson T. Signaling properties of CR3 (CD11b/CD18) and CR1 (CD35) in relation to phagocytosis of complement-opsonized particles. J Immunol 1993; 151:330-8.
107. Widmann C, Gibson S, Jarpe MB et al. Mitogen-activated protein kinase: Conservation of a three-kinase module from yeast to human. Physiol Rev 1999; 79:143-80.

108. Mansfield PJ, Shayman JA, Boxer LA. Regulation of polymorphonuclear leukocyte phagocytosis by myosin light chain kinase after activation of mitogen-activated protein kinase. Blood 2000; 95:2407-2412.
109. Kalhammer G, Bähler M. Unconventional myosins. Essays Biochem 2000; 35:33-42.
110. Reddy SM, Hsiao KH, Abernethy VE et al. Phagocytosis of apoptotic cells by macrophages induces novel signaling events leading to cytokine-independent survival and inhibition of proliferation: Activation of Akt and inhibition of extracellular signal-regulated kinases 1 and 2. J Immunol 2002; 169:702-13.
111. Kurosaka K, Takahashi M, Kobayashi Y. Activation of extracellular signal-regulated kinase 1/2 is involved in production of CXC-chemokine by macrophages during phagocytosis of late apoptotic cells. Biochem Biophys Res Commun 2003; 306:1070-4.
112. Kurosaka K, Watanabe N, Kobayashi Y. Production of proinflammatory cytokines by resident tissue macrophages after phagocytosis of apoptotic cells. Cell Immunol 2001; 211:1-7.
113. Castellano F, Chavrier P, Caron E. Actin dynamics during phagocytosis. Semin Immunol 2001; 13:347-55.
114. Wittmann T, Waterman-Storer CM. Cell motility: Can Rho GTPases and microtubules point the way? J Cell Sci 2001; 114:3795-803.
115. Swanson J, Baer S. Phagocytosis by zippers and triggers. Trends Cell Biol 1995; 5:89-93.
116. Newman SL, Mikus LK, Tucci MA. Differential requirements for cellular cytoskeleton in human macrophage complement receptor- and Fc receptor-mediated phagocytosis. J Immunol 1991; 146:967-74.
117. Welch MD, Mullins RD. Cellular control of actin nucleation. Annu Rev Cell Dev Biol 2002; 18:247-88.
118. Turner M, Billadeau D. VAV proteins as signal integrators for multi-subunit immune-recognition receptors. Nat Rev Immunol 2002; 7:476-86.
119. Dib K, Melander F, Andersson T. Role of p190RhoGAP in beta 2 integrin regulation of RhoA in human neutrophils. J Immunol 2001; 166:6311-22.
120. Patel JC, Hall A, Caron E. Vav regulates activation of Rac but not Cdc42 during FcgammaR-mediated phagocytosis. Mol Biol Cell 2002; 13:1215-26.
121. Coppolino MG, Krause M, Hagendorff P et al. Evidence for a molecular complex consisting of Fyb/SLAP, SLP-76, Nck, VASP and WASP that links the actin cytoskeleton to Fcgamma receptor signalling during phagocytosis. J Cell Sci 2001; 114:4307-18.
122. Zhou X, Li J, Kucik DF. The microtubule cytoskeleton participates in control of beta2 integrin avidity. J Biol Chem 2001; 276:44762-9.
123. Gumienny TL, Brugnera E, Tosello-Trampont AC et al. CED-12/ELMO, a novel member of the CrkII/Dock180/Rac pathway, is required for phagocytosis and cell migration. Cell 2001; 107:27-41.
124. Le Cabec V, Carreno S, Moisand A et al. Complement receptor 3 (CD11b/CD18) mediates type I and type II phagocytosis during nonopsonic and opsonic phagocytosis, respectively. J Immunol 2002; 169:2003-9.
125. Tosello-Trampont A, Nakada-Tsukui K, KS R. Engulfment of apoptotic cells is negatively regulated by Rho-mediated signaling. 2003, (Epub ahead of print Sep 26).
126. May RC, Caron E, Hall A et al. Involvement of the Arp2/3 complex in phagocytosis mediated by FcgR or CR3. Nature Cell Biology 2000; 2:246-248.
127. Olazabal IM, Caron E, May RC et al. Rho-kinase and myosin-II control phagocytic cup formation during CR, but not FcgammaR, phagocytosis. Curr Biol 2002; 12:1413-18.
128. Lorenzi R, Brickell PM, Katz DR et al. Wiskott-Aldrich syndrome protein is necessary for efficient IgG-mediated phagocytosis. Blood 2000; 95:2943-6.
129. Leverrier Y, Lorenzi R, Blundell MP et al. Cutting edge: The Wiskott-Aldrich syndrome protein is required for efficient phagocytosis of apoptotic cells. J Immunol 2001; 166:4831-4.
130. Allen LA, Aderem A. A role for MARCKS, the alpha isozyme of protein kinase C and myosin I in zymosan phagocytosis by macrophages. J Exp Med 1995; 182:829-840.
131. Scott MP, Zappacosta F, Kim EY et al. Identification of novel SH3 domain ligands for the Src family kinase Hck. Wiskott-Aldrich syndrome protein (WASP), WASP-interacting protein (WIP), and ELMO1. J Biol Chem 2002; 277:28238-46.
132. Swanson JA, Johnson MT, Beningo K et al. A contractile activity that closes phagosomes in macrophages. J Cell Sci 1999; 112:307-316.
133. Araki N, Hatae T, Furukawa A et al. Phosphoinositide-3-kinase-independent contractile activities associated with Fcgamma-receptor-mediated phagocytosis and macropinocytosis in macrophages. J Cell Sci 2003; 116:247-57.

CHAPTER 2

Phagocytosis and Immunity

Steven Greenberg

Abstract

Phagocytosis is an phylogenetically conserved mechanism utilized by many cells to ingest microbial pathogens and apoptotic or necrotic corpses. Recent studies have demonstrated that phagocytosis serves to initiate immunity mediated by both Class I and Class II MHC. Depending on the identity of the specific phagocytic receptor involved, phagocytosis can either enhance or suppress inflammation. Dysregulation of phagocytosis can lead to alterations in the immune response and may contribute to autoimmunity. Harnessing the phagocytic capacity of antigen presenting cells may ultimately lead to exploitation of phagocytosis as a therapeutic modality in intractable diseases, such as advanced cancer.

Phagocytosis is the process by which leukocytes and other cells ingest particulate ligands whose size exceeds about 1 µm. This phylogenetically ancient cellular event is critical for both innate and acquired immunity. By ingesting microbial pathogens, phagocytic leukocytes accomplish two essential immune functions. First, they initiate a microbial death pathway. They target ingested pathogens to degradative organelles, such as lysosomes and to vesicles containing components of the phagocyte oxidase complex. Second, phagocytic leukocytes, particularly dendritic cells (DCs), utilize phagocytosis to direct antigens to both MHC I and II compartments. Thus, phagocytosis serves a dual role as an effector of innate immunity and an initiator of acquired immunity.

Diversity of Phagocytic Receptors

Many receptors are capable of mediating phagocytosis (Table 1). These receptors can be broadly defined as those that recognize epitopes on the surfaces of unmodified bacteria and fungi (nonopsonic phagocytosis) or those that recognize components derived from the host (opsonic phagocytosis). Examples of the latter include receptors for the Fc portion of IgG (FcγRs) and receptors that recognize various components of complement (e.g., Complement Receptor 3). It is important to distinguish between those receptors that participate solely in binding of the phagocytic target with receptors that are actually coupled to the transmembrane signaling machinery of phagocytosis. For example, many members of the Scavenger Receptor superfamily are "pattern recognition receptors," which bind ligands, such as peptidoglycan, present on the surfaces of bacteria and fungi.[1] However, most of these receptors have not been shown to trigger distinct transmembrane signals; rather, they facilitate the recruitment of other receptors that directly engage the phagocytic machinery.

Molecular Mechanisms of Phagocytosis, edited by Carlos Rosales. ©2005 Eurekah.com and Springer Science+Business Media.

Table 1. Examples of receptors in mammalian cells that promote phagocytosis[a]

Receptor	Cell Type	Target	Ligand	Refs.
FcγRI FcγRIIA FcγRIII	Gran, Mo, MΦ, DCs, Mast, Plts	IgG-coated pathogens	Fc port ion of IgG	many
CR1 (CD35)	PMN, Mo, MF	Complement-opsonized bacteria and fungi	C3b, C4b Mannan-binding lectin	84
CR3 (CD11b/CD18; $\alpha_M\beta_3$; Mac1)	PMN, Mo, MΦ, myeloid DCs	Complement-opsonized bacteria and fungi Gram-negative bacteria *Bordatella* pertussis Yeast	C3bi, C3d LPS Filamentous hemagglutinin β-glucan	85
CR4 (CD11c/ CD18; p150, 95)	MΦ, DC	*M. tuberculosis*	?	86
CD48	Mast	Enterobacteria	FimH	87
Dectin-1	MΦ	*P. carinii, C. albicans*	Mannosyl/fucosyl residues	88
Scavenger receptor AI/II	MΦ	Apoptotic lymphocytes Gram-positive cocci	?PS Leipoteichoic acid	89-91
Scavenger receptor BI	Sertoli cells, Thymic Epi	Apoptotic cells	PS	92,93
MARCO	MΦ	*E. coli, S. aureus*	?	94
Mer	MΦ	Apoptotic thymocytes	?Gas6/PS	95
PSR	Many	Apoptotic cells	PS	96
CD36	MΦ	Apoptotic PMN	PS/Thrombospondin	97,98
CD14	MΦ	*P. aeruginosa* Apoptotic cells	?LPS ?	99,100
β_1 integrins	Many	*Yersinia*	Invasin	101
$\alpha_v\beta_3$	MΦ	Apoptotic cells	?Thrombospondin	97,102
$\alpha_v\beta_5$	DC, Epi	Apoptotic cells	?	103,104
E-cadherin	Epi	*Listeria*	InlA	34
Met	Epi	*Listeria*	InlB	31

[a]Specific inhibition of binding by these receptors correlates with inhibition of phagocytosis. However, with some notable exceptions (e.g., FcγRIIA and the macrophage mannose receptor), it is possible that the indicated receptor serves to enhance ligand binding, rather than to participate directly in the ingestion process; [b]DC = dendritic cells; Epi = epithelial cells; Leuk = leukocytes; Mast = mast cells; Mo = monocytes; Mφ = macrophages; PMN = polymorphonuclear leukocytes; PS = phosphatidylserine

Mechanisms of Phagocytic Signaling

The signaling mechanisms of phagocytosis are well-understood for only a handful of receptors.[2,3] FcγRs signal by recruiting an array of tyrosine kinases in the vicinity of the ligated receptors. FcγR ligand binding receptors or their subunits contain immunoreceptor tyrosine-based activating motif (ITAMs) in their cytosolic domains. These motifs become phosphorylated by members of the Src family,[4] which serve as high-affinity binding sites for SH2 domains of Syk, a tyrosine kinase that is expressed predominantly in hematopoietic cells. Although Syk is clearly required for phagocytosis,[5-8] the identity of further downstream signals that are critical for phagocytosis are less certain. Phospholipid kinases (phosphatidylinositol 3-kinase; PI 3-kinase[9-11] and phosphatidylinositol-4-phosphate 5-kinase[12]) are clearly involved as are various serine/threonine protein kinases (e.g., MEK1/2 and/or ERK, in the case of neutrophils[13]) and PKC.[14] PLA$_2$ and PLD are also activated and believed to participate in the phagocytic process.[15,16] The former may participate in vesicle trafficking during phagocytosis[17] as well as contribute to the production of leukotrienes that amplify the phagocytic signal.[18] Ultimately, early phagocytic signaling events culminate in net actin assembly and pseudopod extension. Actin assembly during phagocytosis is mediated by one or more Rho family GTPases.[19-22] Other GTPases may participate in actin polymerization and/or vesicle trafficking, such as ARF6.[23,24] In contrast, PI 3-kinase regulates pseudopod extension,[9] in part by recruiting intracellular pools of latent phagosomal membrane,[25,26] and in part by recruiting PH domain-containing proteins, such as myosin-X, to the phagosome.[27]

The occurence of ITAMs in other phagocytic receptors suggests strongly that the paradigm of phagocytic signaling utilized by FcγRs is likely to be a general one. For example, Dectin-1, a lectin that recognizes mannosyl/fucosyl residues on fungi,[28] and CEACAM3, a receptor that recognizes on *Neisseria*, *Moraxella*, and *Haemophilus* species contain functional ITAMs in their cytosolic domains.[29,30]

Fate of Engulfed Pathogens—Us vs. Them

The initial host response to most bacterial and fungal pathogens is phagocytosis. The particular route of entry is a function of the nature of the pathogen being ingested and identity of the host cell receptors engaging the pathogen. For example, internalization of *Listeria* is mediated by the adhesins InlA, which binds to E-cadherin on host epithelia, and InlB, which binds to the Met tyrosine kinase and to gC1q-R on host cells.[31,32] E-cadherin-mediated entry requires participation of catenins[33,34] and Met-dependent signaling induces activation of PI 3-kinase. For *Yersinia*, recognition of invasin on the bacterial surface is mediated by β$_1$ integrins on a variety of cells; bacterial uptake requires the participation of Src-family tyrosine kinases and focal adhesion kinase.[35]

Phagocytosis can be a highly localized event (e.g., phagocytosis mediated by "zippers") or can be partially delocalized by virtue of a diffusible mediator. Several pathogens, such as *Salmonella* or *Shigella*, stimulate a "trigger" mechanism of invasion, inducing the assembly of actin in the host cell in the vicinity of where the bacteria interact with the host.[36] Using a Type III secretion system, Salmonella injects SopE, a protein with guanine nucleotide exchange factor for Cdc42 and Rac, into host cells.[37] In contrast, some pathogens, such as *Haemophilus ducreyi*[38] and *Yersinia*, use different strategies to evade phagocytosis. *Yersinia* secretes YopH, a tyrosine phosphatase that dephosphorylates the focal adhesion protein, Cas, as well as other potential tyrosine kinase substrates.[39] Another secreted product of *Yersinia*, YopE, is a RhoGAP.[40]

Evolutionary pressure results in genetically stable adaptations by bacteria that serve to compromise the host. Many pathogens evade killing by inducing a delay in phagosome maturation. This can result in exclusion of active lysosomal enzymes or components of the NADPH-oxidase-containing vesicles from the phagosomes. Among the survival strategies employed by *Mycobacterium tuberculosis*, for example, is the suppression of calcium signaling, which contributes to evasion of lysosome fusion by the *Mycobacterium*-containing phagosome.[41] Some of the phagosome maturation arrest activity resides in glycosylated lipids derived from the cell wall of *M. tuberculosis*.[42-44] Other pathogens, such as *Legionella*, avoid maturation at early stages of phagosome biogenesis.[45] The *Legionella* phagocytic vacuole is highly specialized: it does not fuse with lysosomes, fails to acidify,[46] and intercepts vesicular traffic from ER exit sites to create an organelle that permits intracellular replication.[47,48]

Phagocytosis and the MHC Class II Pathway

Once ingested, microbes that reside in phagocytic vacuoles find themselves in an increasingly hostile environment. Phagosomes undergo a maturation process, beginning with recruitment of Rab-5-positive early endosomes, followed by fusion with Rab-7-positive late endosomes.[49] This is accompanied by further maturation and fusion with lysosomes. The capacity of lysosomes to degrade proteins is under developmental control. Mature dendritic cells demonstrate an enhanced capacity to degrade antigen, which correlates with a greater acidification of lysosomes and enhanced lysosomal vacuolar-ATPase activity.[50] Following degradation and loading onto MHC Class II, mature DCs generate tubules from lysosomal compartments, which fuse directly with the plasma membrane.[50]

Phagocytosis and the MHC Class I Pathway-Intracellular Pathogens and Tumors

In the past few years, much progress has been made to explain the cellular basis for the phenomenon termed "cross-priming," first described by Bevan in 1976.[51] This method of antigen presentation relies on the phagocytosis of an apoptotic cell (e.g., induced by viral infection) and the presentation of phagocytically-derived viral antigens onto MHC Class I. Previously, it had been thought that loading onto MHC Class I occurs only following loading of endogenous antigen from the cytosol to the ER. In contrast, the "phagosome-to-cytosol" pathway of antigen processing involves translocation of peptides or proteins from within the phagosome to the cytoplasm.[52] There is precedence for proteins crossing membrane barriers; this is the principal mechanism for protein import into mitochondria, for example. How do phagosomal membranes become modified to accomplish a similar task? The observation that ER membrane has the capacity to fuse with phagosomal membrane offers one potential mechanism for this mode of antigen presentation.[53,54] Sec61, an ER-derived protein translocation channel, was observed to become incorporated into phagosomes of dendritic cells.[54] Because this protein is capable of "reverse transport" of proteins (in the opposite direction of its recognized ability to translocate nascent chains into the ER),[55] it is possible that Sec61 provides the means by which proteins translocate across the membrane of the "phago-ER-some" into the cytosol. Once in the cytosol, the protein can be targeted to the proteasome for degradation and further MHC Class I processing by the conventional TAP-dependent pathway.

Cross-priming may be essential for immunity to viruses and other intracellular pathogens. Using a mouse model of viral immunity, Rock and colleagues demonstrated that virally infected nonhematopoietic cells are unable to stimulate primary CTL-mediated immunity directly.[56] Instead, bone-marrow-derived cells are required as antigen-presenting cells to initiate anti-viral CTL responses. From a teleological standpoint this makes sense. Viruses that typically do not infect hematopoietic cells fail to gain entrance to secondary lymphoid organs. Therefore, they are unlikely to serve as efficient antigen-presenting cells in a primary immune response. However, this interpretation has been called into question. In one recent study, poliovirus in naturally nonpermissive murine APCs acquired viral RNA in vivo independently of the cellular virus receptor. The polioviral RNA initiated neosynthesis of viral antigen sufficient to prime CTLs in vivo.[57] It thus remains an open question as to the relative importance of the endogenous pathway and cross-priming in MHC Class I-restricted immunity in vivo.

Cross-priming may be critical for tumor immunity. Although this pathway has been demonstrated in a variety of mononuclear phagocytes, it appears that dendritic cells are the most potent APCs in stimulating phagosome-derived MHC Class I-restricted CD8+ CTLs.[58] Many strategies for using dendritic cells loaded with tumor-derived antigens ex vivo have been proposed. Effective tumor immunity requires efficient loading of antigen and effective targeting of antigen-loaded DCs to reach secondary lymphoid organs. As this normally requires a maturation stimulus for the antigen-loaded DCs, various stimuli that serve as adjuvants for the anti-tumor response are under investigation.[59-61] These include tumor-derived heat-shock proteins,[62] pro-inflammatory cytokines, and direct activation of costimulatory molecules on

DCs.[63,64] In some cases, antigen-loading and maturation stimuli are triggered by the same phagocytically-competent receptor. This is the case for Fcγ receptors and complexes of tumor antigen and IgG.[65,66]

Ingestion of Dead Cells and Regulation of Inflammation; The Power of "Negative Signaling"

Apoptotic cell death is a consequence of cellular senescence or infection of cells with various microbial pathogens. Effective removal of these cells is required for tissue remodeling, for cross-priming, and for resolution of inflammation. This occurs by phagocytosis, which accomplishes the dual role of facilitating the clearance of apoptotic bodies and completing the cell death pathway.[67] Many receptors on the phagocyte surface participate in this process; among these is the recently identified phosphatidylserine receptor (PSR), which recognizes PS exposed at the outer leaflet of the membrane of apoptotic cells and triggers the generation of TGF-β by the phagocyte. Recognition of apoptotic cells by phagocytes is enhanced by chemotactic stimuli released by apoptotic cells, such as lysophosphatidylcholine.[68]

The consequences of ingestion of dead cells depends on the mechanism of cell death. Necrotic cells release a variety of pro-inflammatory substances, including heat shock proteins, which engage a subset of receptors on the phagocyte surfaces. These, in turn, bind cell surface proteins such as CD91,[69] which potentially mediates endocytosis and cross-presentation. Heat shock proteins have been proposed to serve as adjuvants by binding toll-like receptors,[70] although recent work has ascribed this function to contaminating LPS.[71] In contrast, engagement of the PSR results in the production of TGF-β, which serves to dampen inflammation, especially mediated by macrophages.[72] This may be critical in tuning the immune response; too much inflammation at an inappropriate time may result in excessive antigen presentation and, theoretically, autoimmunity (see below).

Another trigger of "negative signaling" is ligation of immunoreceptor tyrosine-based inhibitory motif (ITIM)-containing receptors. In contrast to Fcγ receptors I, IIa, and III, FcγRIIb contains an ITIM in its cytosolic domain. Coclustering of FcγRIIb with any other receptor results in an inhibition of phagocytosis and inflammation by recruitment of the SH2 domain-containing inositol 5' phosphatase (SHIP). SHIP hydrolyzes the lipid product of PI 3-kinase, phosphatidylinositol-3,4,5-trisphosphate to phosphatidylinositol-3,4-bisphosphate, thereby limiting recruitment and activation of several PH domain-containing proteins, such as members of the Tec family of tyrosine kinases. These enzymes phosphorylate and activate phospholipase C-γ; therefore, recruitment of SHIP via ITIMs effectively inhibits signaling mediated by phospholipase C-γ, including calcium fluxes. Recruitment of SHIP via of FcγRIIb may be critical for the maintenance of tolerance as FcγRIIb[-/-] mice develop spontaneous autoimmunity in a B cell-autonomous fashion.[73] Another type of ITIM bearing receptor, SIRPa, is expressed in myeloid cells and recruits the tyrosine phosphatase SHP-1 and SHP-2. One of its ligands, Surfactant Protein-A,[74] is also an opsonin for apoptotic cells.[75] Thus, apoptotic cells may generate mechanistically distinct inhibitory signals: production of TGF-β, a diffusible mediator, via the PSR, and recruitment of the cell autonomous inhibitor, SHP-1/2, by engagement of SIRPa. In other scenarios, SIRPa may serve as an "anti-phagocytic" receptor.[76] For example, the presence of its counter-receptor on red blood cells, CD47, contributes to their lack of recognition as a phagocytic target. It has been suggested that failure of cognate interactions between CD47 on red blood cells and SIRPa on phagocytic cells may contribute to autoimmune hemolytic anemia.[77]

In summary, by recruitment of enzymes that acts to antagonize tyrosine kinase-mediated signaling (SHP-1/SHP-2) or PI 3-kinase-mediated signaling (SHIP), or by generation of TGF-β, the program of inflammation that accompanies phagocytosis can be curtailed or potentially reversed.

Phagocytosis and Autoimmunity

Individuals with various complement deficiences are at greatly increased risk for developing autoimmune diseases.[78] Interestingly, these same complement components are recognized opsonins for apoptotic cells. It has been suggested that decreased clearance of apoptotic bodies leads to an abundant source of potential autoantigens and predisposes to autoimmunity.[79] This model is supported by the observation of anti-DNA antibodies in mice deficient in opsonizing complement components. In the case of C1q-deficient mice, this was accompanied by enhanced accumulation of apoptotic bodies, suggesting that C1q deficiency causes autoimmunity by impairment of the clearance of apoptotic cells.[80] A similar correlation between delayed clearance of apoptotic cells and autoimmunity was observed in mice that lacked functional c-Mer, a tyrosine kinase that engages apoptotic cells and mediates their phagocytosis.[81] Thus, the development of autoimmunity is associated with a defect in the phagocytic clearance of apoptotic cells. Despite the appeal of this interpretation, it is not known whether autoimmunity and defective phagocytosis are causally related. An alternative, though not mutually exclusive explanation, is that complement is required for presentation of self-antigen, such as chromatin, by complement receptor-bearing stromal cells in the bone marrow. These are presented to potentially immature autoreactive B cells leading to tolerization through clonal deletion and/or anergy.[82] It is also possible that lack of appropriate phagocytic clearance of apoptotic cells results in an imbalance of pro- and anti-inflammatory signals, favoring the former. While this seems counterintuitive, as the "anti-inflammatory" PSR pathway of apoptotic cell clearance would be preserved under these circumstances, other anti-inflammatory pathways of corpse clearance may exist. For example, the acute phase reactant C-reactive protein binds to apoptotic cells and serves as both an opsonin and an inducer of TGF-β in a C1q-dependent fashion.[83]

Conclusions

The study of phagocytosis and its assigned role in immunity has come a long way since Metchnikoff. Phagocytosis is now viewed as a fundamental mechanism of antigen presentation, both through MHC Class II and Class I (via cross-priming). This has implications for immunity not only to bacteria and fungi, but also to viruses and malignant cells. The dichotomy of "pro-inflammatory" and "anti-inflammatory" phagocytosis is useful in explaining how the immune system regulates the temporal response to infection. The balance between these pathways may prove to be a key determinant of the development of autoimmunity.

Acknowledgements

I thank Benjamin Dale and Peter Henson for their critical reading of the manuscript. This work was supported by NIH HL54164.

References

1. Franc NC, White K, Ezekowitz RAB. Phagocytosis and development: Back to the future. Curr Opin Immunol 1999; 11(1):47-52.
2. Greenberg S, Grinstein S. Phagocytosis and innate immunity. Curr Opin Immunol 2002; 14(1):136-145.
3. Garcia-Garcia E, Rosales C. Signal transduction during Fc receptor-mediated phagocytosis. J Leukoc Biol 2002; 72(6):1092-1108.
4. Fitzer-Attas CJ, Lowry M, Crowley MT et al. Fcγ receptor-mediated phagocytosis in macrophages lacking the Src family tyrosine kinases Hck, Fgr, and Lyn. J Exp Med 2000; 191(4):669-682.
5. Crowley MT, Costello PS, Fitzer-Attas CJ et al. A critical role for Syk in signal transduction and phagocytosis mediated by Fcγ receptors on macrophages. J Exp Med 1997; 186:1027-1039.
6. Cox D, Chang P, Kurosaki T et al. Syk tyrosine kinase is required for immunoreceptor tyrosine activation motif-dependent actin assembly. J Biol Chem 1996; 271(28):16597-16602.

7. Matsuda M, Park JG, Wang DC et al. Abrogation of the Fcγ receptor IIA-mediated phagocytic signal by stem-loop Syk antisense oligonucleotides. Mol Biol Cell 1996; 7(7):1095-1106.

8. Kiefer F, Brumell J, Al-Alawi N et al. The Syk protein tyrosine kinase is essential for Fcγ receptor signaling in macrophages and neutrophils. Mol Cell Biol 1998; 18(7):4209-4220.

9. Cox D, Tseng C-C, Bjekic G et al. A requirement for phosphatidylinositol 3-kinase in pseudopod extension. J Biol Chem 1999; 274:1240-1247.

10. Ninomiya N, Hazeki K, Fukui Y et al. Involvement of phosphatidylinositol 3-kinase in Fcγ receptor signaling. J Biol Chem 1994; 269:22732-22737.

11. Araki N, Johnson MT, Swanson JA. A role for phosphoinositide 3-kinase in the completion of macropinocytosis and phagocytosis by macrophages. J Cell Biol 1996; 135:1249-1260.

12. Botelho RJ, Teruel M, Dierckman R et al. Localized biphasic changes in phosphatidylinositol-4,5-bisphosphate at sites of phagocytosis. J Cell Biol 2000; 151(7):1353-1368.

13. Mansfield PJ, Shayman JA, Boxer LA. Regulation of polymorphonuclear leukocyte phagocytosis by myosin light chain kinase after activation of mitogen-activated protein kinase. Blood 2000; 95(7):2407-2412.

14. Larsen EC, DiGennaro JA, Saito N et al. Differential requirement for classic and novel PKC isoforms in respiratory burst and phagocytosis in RAW 264.7 cells. J Immunol 2000; 165(5):2809-2817.

15. Lennartz MR, Brown EJ. Arachidonic acid is essential for IgG Fc receptor-mediated phagocytosis by human monocytes. Journal of Immunology 1991; 147(2):621-626.

16. Kusner DJ, Hall CF, Jackson S. Fcγ receptor-mediated activation of phospholipase D regulates macrophage phagocytosis of IgG-opsonized particles. J Immunol 1999; 162:2266-2274.

17. Lennartz MR, Yuen AFC, Masi SM et al. Phospholipase A₂ inhibition results in sequestration of plasma membrane into electron-lucent vesicles during IgG-mediated phagocytosis. J Cell Sci 1997; 110:2041-2052.

18. Mancuso P, Nana-Sinkam P, Peters-Golden M. Leukotriene B₄ augments neutrophil phagocytosis of Klebsiella pneumoniae. Infect Immun 2001; 69(4):2011-2016.

19. Cox D, Chang P, Zhang Q et al. Requirements for both Rac1 and Cdc42 in membrane ruffling and phagocytosis in leukocytes. J Exp Med 1997; 186:1487-1494.

20. Forsberg M, Druid P, Zheng L et al. Activation of Rac2 and Cdc42 on Fc and complement receptor ligation in human neutrophils. J Leukoc Biol 2003; 74(4):611-619, Epub 2003 Jul 2001.

21. Caron E, Hall A. Identification of two distinct mechanisms of phagocytosis controlled by different Rho GTPases. Science 1998; 282:1717-1721.

22. Massol P, Montcourrier P, Guillemot J-C et al. Fc receptor-mediated phagocytosis requires CDC42 and Rac1. EMBO J 1998; 17:6219-6229.

23. Zhang Q, Cox D, Tseng C-C et al. A requirement for ARF6 in Fcγ receptor-mediated phagocytosis in macrophages. J Biol Chem 1998; 273:19977-19981.

24. Niedergang F, Colucci-Guyon E, Dubois T et al. ADP ribosylation factor 6 is activated and controls membrane delivery during phagocytosis in macrophages. J Cell Biol 2003; 161(6):1143-1150, Epub 2003 Jun 1116.

25. Cox D, Lee DJ, Dale BM et al. A Rab11-containing rapidly recycling compartment in macrophages that promotes phagocytosis. Proc Natl Acad Sci USA 2000; 97:680-685.

26. Hackam DJ, Rotstein OD, Sjolin C et al. v-SNAREdependent secretion is required for phagocytosis. Proc Natl Acad Sci USA 1998; 95(20):11691-11696.

27. Cox D, Berg JS, Cammer M et al. Myosin X is a downstream effector of PI(3)K during phagocytosis. Nature Cell Biology 2002; 4(7):469-477.

28. Steele C, Marrero L, Swain S et al. Alveolar macrophage-mediated killing of Pneumocystis carinii f. sp. muris involves molecular recognition by the Dectin-1 β-glucan receptor. J Exp Med 2003; 198(11):1677-1688.

29. Schmitter T, Agerer F, Peterson L et al. Granulocyte CEACAM3 Is a phagocytic receptor of the innate immune system that mediates recognition and elimination of human-specific pathogens. J Exp Med 2004; 199(1):35-46.

30. McCaw SE, Schneider J, Liao EH et al. Immunoreceptor tyrosine-based activation motif phosphorylation during engulfment of Neisseria gonorrhoeae by the neutrophil-restricted CEACAM3 (CD66d) receptor. Mol Microbiol 2003; 49(3):623-637.

31. Shen Y, Naujokas M, Park M et al. InlB-dependent internalization of Listeria is mediated by the Met receptor tyrosine kinase. Cell 2000; 103(3):501-510.

32. Braun L, Ghebrehiwet B, Cossart P. gC1q-R/p32, a C1q-binding protein, is a receptor for the InlB invasion protein of Listeria monocytogenes. EMBO J 2000; 19(7):1458-1466.
33. Lecuit M, Hurme R, Pizarro-Cerda J et al. A role for α-and β-catenins in bacterial uptake. Proc Natl Acad Sci USA 2000; 97(18):10008-10013.
34. Mengaud J, Ohayon H, Gounon P et al. E-cadherin is the receptor for internalin, a surface protein required for entry of L. monocytogenes into epithelial cells. Cell 1996; 84(6):923-932.
35. Alrutz MA, Isberg RR. Involvement of focal adhesion kinase in invasin-mediated uptake. Proc Natl Acad Sci USA 1998; 95(23):13658-13663.
36. Tran Van Nhieu G, Caron E, Hall A et al. IpaC induces actin polymerization and filopodia formation during Shigella entry into epithelial cells. EMBO J 1999; 18(12):3249-3262.
37. Hardt W-D, Chen L-M, Schuebel KE et al. S. typimurium encodes an activator of Rho GTPases that induces membrane ruffling and nuclear responses in host cells. Cell 1998; 93:815-826.
38. Vakevainen M, Greenberg S, Hansen EJ. Inhibition of phagocytosis by Haemophilus ducreyi requires expression of the LspA1 and LspA2 proteins. Infect Immun 2003; 71(10):5994-6003.
39. Black DS, Bliska JB. Identification of p130^Cas as a substrate of Yersinia YopH (Yop51), a bacterial protein tyrosine phosphatase that translocates into mammalian cells and targets focal adhesions. EMBO J 1997; 16(10):2730-2744.
40. Black DS, Bliska JB. The RhoGAP activity of the Yersinia pseudotuberculosis cytotoxin YopE is required for antiphagocytic function and virulence. Mol Microbiol 2000; 37(3):515-527.
41. Malik ZA, Denning GM, Kusner DJ. Inhibition of Ca^{2+} signaling by Mycobacterium tuberculosis is associated with reduced phagosome-lysosome fusion and increased survival within human macrophages. J Exp Med 2000; 191(2):287-302.
42. Fratti RA, Chua J, Vergne I et al. Mycobacterium tuberculosis glycosylated phosphatidylinositol causes phagosome maturation arrest. Proc Natl Acad Sci USA 2003; 100(9):5437-5442, Epub 2003 Apr 5417.
43. Russell DG. Phagosomes, fatty acids and tuberculosis. Nat Cell Biol 2003; 5(9):776-778.
44. Anes E, Kuhnel MP, Bos E et al. Selected lipids activate phagosome actin assembly and maturation resulting in killing of pathogenic mycobacteria. Nat Cell Biol 2003; 5(9):793-802, Epub 2003 Aug 2024.
45. Clemens DL, Lee BY, Horwitz MA. Deviant expression of Rab5 on phagosomes containing the intracellular pathogens Mycobacterium tuberculosis and Legionella pneumophila is associated with altered phagosomal fate. Infec Immunity 2000; 68(5):2671-2684.
46. Horwitz MA, Maxfield FR. Legionella pneumophila inhibits acidification of its phagosome in human moncytes. J Cell Biol 1984; 99:1936-1943.
47. Horwitz M. Phagocytosis of the Legionnaires' disease bacterium (Legionella pneumophila) occurs by a novel mechanism: Engulfment within a pseudopod coil. Cell 1984; 36:27-33.
48. Kagan JC, Roy CR. Legionella phagosomes intercept vesicular traffic from endoplasmic reticulum exit sites. Nat Cell Biol 2002; 4(12):945-954.
49. Harrison RE, Bucci C, Vieira OV et al. Phagosomes fuse with late endosomes and/or lysosomes by extension of membrane protrusions along microtubules: Role of Rab7 and RILP. Mol Cell Biol 2003; 23(18):6494-6506.
50. Trombetta ES, Ebersold M, Garrett W et al. Activation of lysosomal function during dendritic cell maturation. Science 2003; 299(5611):1400-1403.
51. Bevan MJ. Cross-priming for a secondary cytotoxic response to minor H antigens with H-2 congenic cells which do not cross-react in the cytotoxic assay. J Exp Med 1976; 143(5):1283-1288.
52. Thery C, Amigorena S. The cell biology of antigen presentation in dendritic cells. Curr Opin Immunol 2001; 13(1):45-51.
53. Guermonprez P, Saveanu L, Kleijmeer M et al. ER-phagosome fusion defines an MHC Class I cross-presentation compartment in dendritic cells. Nature 2003; 425(6956):397-402.
54. Houde M, Bertholet S, Gagnon E et al. Phagosomes are competent organelles for antigen cross-presentation. Nature 2003; 425(6956):402-406.
55. Wiertz EJ, Tortorella D, Bogyo M et al. Sec61-mediated transfer of a membrane protein from the endoplasmic reticulum to the proteasome for destruction. Nature 1996; 384(6608):432-438.
56. Sigal LJ, Crotty S, Andino R et al. Cytotoxic T-cell immunity to virus-infected nonhaematopoietic cells requires presentation of exogenous antigen. Nature 1999; 398(6722):77-80.

57. Freigang S, Egger D, Bienz K et al. Endogenous neosynthesis vs. Cross-presentation of viral antigens for cytotoxic T cell priming. Proc Natl Acad Sci USA 2003; 100(23):13477-13482, Epub 2003 Oct 13431.
58. Albert ML, Sauter B, Bhardwaj N. Dendritic cells acquire antigen from apoptotic cells and induce class I- restricted CTLs. Nature 1998; 392(6671):86-89.
59. Ochsenbein AF. Principles of tumor immunosurveillance and implications for immunotherapy. Cancer Gene Ther 2002; 9(12):1043-1055.
60. Arina A, Tirapu I, Alfaro C et al. Clinical implications of antigen transfer mechanisms from malignant to dendritic cells. Exploiting cross-priming. Exp Hematol 2002; 30(12):1355-1364.
61. Ward S, Casey D, Labarthe MC et al. Immunotherapeutic potential of whole tumour cells. Cancer Immunol Immunother 2002; 51(7):351-357, Epub 2002 Jun 2014.
62. Suto R, Srivastava PK. A mechanism for the specific immunogenicity of heat shock protein-chaperoned peptides. Science 1995; 269(5230):1585-1588.
63. Mackey MF, Gunn JR, Maliszewsky C et al. Dendritic cells require maturation via CD40 to generate protective antitumor immunity. J Immunol 1998; 161(5):2094-2098.
64. Labeur MS, Roters B, Pers B et al. Generation of tumor immunity by bone marrow-derived dendritic cells correlates with dendritic cell maturation stage. J Immunol 1999; 162(1):168-175.
65. Regnault A, Lankar D, Lacabanne V et al. Fcγ receptor-mediated induction of dendritic cell maturation and major histocompatibility complex class I-restricted antigen presentation after immune complex internalization. J Exp Med 1999; 189(2):371-380.
66. Rafiq K, Bergtold A, Clynes R. Immune complex-mediated antigen presentation induces tumor immunity. J Clin Invest 2002; 110(1):71-79.
67. Hoeppner DJ, Hengartner MO, Schnabel R. Engulfment genes cooperate with ced-3 to promote cell death in Caenorhabditis elegans. Nature 2001; 412(6843):202-206.
68. Lauber K, Bohn E, Krober SM et al. Apoptotic cells induce migration of phagocytes via caspase-3-mediated release of a lipid attraction signal. Cell 2003; 113(6):717-730.
69. Binder RJ, Han DK, Srivastava PK. CD91: A receptor for heat shock protein gp96. Nat Immunol 2000; 1(2):151-155.
70. Vabulas RM, Wagner H, Schild H. Heat shock proteins as ligands of toll-like receptors. Curr Top Microbiol Immunol 2002; 270:169-184.
71. Gao B, Tsan MF. Endotoxin contamination in recombinant human heat shock protein 70 (Hsp70) preparation is responsible for the induction of tumor necrosis factor alpha release by murine macrophages. J Biol Chem 2003; 278(1):174-179.
72. Fadok VA, Bratton DL, Konowal A et al. Macrophages that have ingested apoptotic cells in vitro inhibit proinflammatory cytokine production through autocrine/paracrine mechanisms involving TGF-β, PGE2, and PAF. J Clin Invest 1998; 101(4):890-898.
73. Bolland S, Ravetch JV. Spontaneous autoimmune disease in FcγRIIB-deficient mice results from strain-specific epistasis. Immunity 2000; 13(2):277-285.
74. Gardai SJ, Xiao YQ, Dickinson M et al. By binding SIRPα or calreticulin/CD91, lung collectins act as dual function surveillance molecules to suppress or enhance inflammation. Cell 2003; 115(1):13-23.
75. Vandivier RW, Ogden CA, Fadok VA et al. Role of surfactant proteins A, D, and C1q in the clearance of apoptotic cells in vivo and in vitro: Calreticulin and CD91 as a common collectin receptor complex. J Immunol 2002; 169(7):3978-3986.
76. Oldenborg PA, Gresham HD, Lindberg FP. CD47-signal regulatory protein alpha (SIRP alpha) regulates Fc gamma and complement receptor-mediated phagocytosis. J Exp Med 2001; 193(7):855-861.
77. Oldenborg PA, Zheleznyak A, Fang YF et al. Role of CD47 as a marker of self on red blood cells. Science 2000; 288(5473):2051-2054.
78. Hart SP, Smith JR, Dransfield I. Phagocytosis of opsonized apoptotic cells: Roles for 'old-fashioned' receptors for antibody and complement. Clin Exp Immunol 2004; 135(2):181-185.
79. Taylor PR, Carugati A, Fadok VA et al. A hierarchical role for classical pathway complement proteins in the clearance of apoptotic cells in vivo. J Exp Med 2000; 192(3):359-366.
80. Bickerstaff MC, Botto M, Hutchinson WL et al. Serum amyloid P component controls chromatin degradation and prevents antinuclear autoimmunity. Nat Med 1999; 5(6):694-697.

81. Cohen PL, Caricchio R, Abraham V et al. Delayed apoptotic cell clearance and lupus-like autoimmunity in mice lacking the c-mer membrane tyrosine kinase. J Exp Med 2002; 196(1):135-140.

82. Paul E, Carroll MC. SAP-less chromatin triggers systemic lupus erythematosus. Nat Med 1999; 5(6):607-608.

83. Gershov D, Kim S, Brot N et al. C-reactive protein binds to apoptotic cells, protects the cells from assembly of the terminal complement components, and sustains an antiinflammatory innate immune response: Implications for systemic autoimmunity. J Exp Med 2000; 192(9):1353-1363.

84. Ghiran I, Barbashov SF, Klickstein LB et al. Complement receptor 1/CD35 is a receptor for mannan-binding lectin. J Exp Med 2000; 192(12):1797-1808.

85. Ross GD. Regulation of the adhesion versus cytotoxic functions of the Mac- 1/CR3/αMβ2-integrin glycoprotein. Crit Rev Immunol 2000; 20(3):197-222.

86. Zaffran Y, Zhang L, Ellner JJ. Role of CR4 in Mycobacterium tuberculosis-human macrophages binding and signal transduction in the absence of serum. Infect Immun 1998; 66(9):4541-4544.

87. Malaviya R, Gao Z, Thankavel K et al. The mast cell tumor necrosis factor alpha response to FimH-expressing Escherichia coli is mediated by the glycosylphosphatidylinositol-anchored molecule CD48. Proc Natl Acad Sci USA 1999; 96(14):8110-8115.

88. Linehan SA, Martinez-Pomares L, Gordon S. Macrophage lectins in host defence. Microbes Infect 2000; 2(3):279-288.

89. Suzuki H, Kurihara Y, Takeya M et al. A role for macrophage scavenger receptors in atherosclerosis and susceptibility to infection. Nature 1997; 386(6622):292-296.

90. Platt N, Suzuki H, Kurihara Y et al. Role for the class A macrophage scavenger receptor in the phagocytosis of apoptotic thymocytes in vitro. Proc Natl Acad Sci USA 1996; 93(22):12456-12460.

91. Thomas CA, Li Y, Kodama T et al. Protection from lethal gram-positive infection by macrophage scavenger receptor-dependent phagocytosis. J Exp Med 2000; 191(1):147-156.

92. Shiratsuchi A, Kawasaki Y, Ikemoto M et al. Role of class B scavenger receptor type I in phagocytosis of apoptotic rat spermatogenic cells by Sertoli cells. J Biol Chem 1999; 274(9):5901-5908.

93. Imachi H, Murao K, Hiramine C et al. Human scavenger receptor B1 is involved in recognition of apoptotic thymocytes by thymic nurse cells. Lab Invest 2000; 80(2):263-270.

94. Palecanda A, Paulauskis J, Al-Mutairi E et al. Role of the scavenger receptor MARCO in alveolar macrophage binding of unopsonized environmental particles. J Exp Med 1999; 189(9):1497-1506.

95. Scott RS, McMahon EJ, Pop SM et al. Phagocytosis and clearance of apoptotic cells is mediated by MER. Nature 2001; 411(6834):207-211.

96. Fadok VA, Bratton DL, Rose DM et al. A receptor for phosphatidylserine-specific clearance of apoptotic cells. Nature 2000; 405(6782):85-90.

97. Savill J, Hogg N, Ren Y et al. Thrombospondin cooperates with CD36 and the vitronectin receptor in macrophage recognition of neutrophils undergoing apoptosis. Journal of Clinical Investigation 1992; 90(4):1513-1522.

98. Fadok VA, Warner ML, Bratton DL et al. CD36 is required for phagocytosis of apoptotic cells by human macrophages that use either a phosphatidylserine receptor or the vitronectin receptor αvβ3. J Immunol 1998; 161(11):6250-6257.

99. Heale J-P, Pollard AJ, Crookall K et al. Two distinct receptors mediate nonopsonic phagocytosis of different strains of Pseudomonas aeruginosa. J Infec Dis 2001; 183(8):1214-1220.

100. Devitt A, Moffatt OD, Raykundalia C et al. Human CD14 mediates recognition and phagocytosis of apoptotic cells. Nature 1998; 392(6675):505-509.

101. Isberg RR, Leong JM. Multiple β$_1$ chain integrins are receptors for invasin, a protein that promotes bacterial penetration into mammalian cells. Cell 1990; 60(5):861-871.

102. Savill J, Dransfield I, Hogg N et al. Vitronectin receptor-mediated phagocytosis of cells undergoing apoptosis. Nature 1990; 343:170-173.

103. Albert ML, Pearce SF, Francisco LM et al. Immature dendritic cells phagocytose apoptotic cells via α$_v$β$_5$ and CD36, and cross-present antigens to cytotoxic T lymphocytes. J Exp Med 1998; 188(7):1359-1368.

104. Finnemann SC, Rodriguez-Boulan E. Macrophage and retinal pigment epithelium phagocytosis: Apoptotic cells and photoreceptors compete for αvβ3 and αvβ5 integrins, and protein kinase C regulates αvβ5 binding and cytoskeletal linkage. J Exp Med 1999; 190(6):861-874.

CHAPTER 3

Fc Receptor Phagocytosis

Randall G. Worth and Alan D. Schreiber

Antigen recognition by cells of the immune system occurs via many mechanisms. One important family of receptors involved in the recognition of immunoglobulin (Ig) coated particles and complexes are Fc receptors. Fc receptors recognize the Fc portion of Ig and are accordingly grouped into subfamilies. They are named depending upon which class of Ig they bind. The major Fc receptors are Fcγ receptors which bind IgG, FcαR that bind IgA and FcεR bind IgE.[1-4] Fc receptors are responsible for such functions as endocytosis,[5-8] phagocytosis,[9-12] granule release,[13-17] reactive mediator release and cell activation/cytotoxicity.[18-23] Fc receptors are found on specific cell types corresponding to their ability to recognize Ig. For example, Fcε receptors are found primarily on eosinophils, basophils and mast cells where they trigger histamine release from intracellular granules whereas Fcγ receptors are found primarily on neutrophils, macrophages and monocytes where they can detect and phagocytose IgG coated pathogens.[14,15]

Fc Receptor Structure

Fc receptors can be divided into two groups based on their signaling capabilities and structure. The first and most common type of Fc receptor is a multi-chain heterocomplex composed of a ligand binding α-chain and one or more signal transducing γ-chains. The second type of Fc receptor is a single-chain transmembrane receptor containing a signal generating motif(s) in the cytoplasmic domain and, thus, not requiring another signal transducing subunit.

In this presentation, the Fcγ receptor family will be discussed as a model system in phagocytic signaling. Fc receptors for a specific isotype of Ig can vary in structure and can differ in ligand affinity and signaling ability (reviewed in Ravetch and Bolland).[24] Fcγ receptors are classified into three classes: FcγRI, FcγRII, and FcγRIII. FcγRI is a high affinity receptor that can be subdivided into three groups (A, B and C) that are encoded by three different genes. Similar to FcγRI, FcγRII has been divided into three families encoded by three genes, also named A, B and C and which are of relatively low affinity for monomeric IgG but of high affinity for complexed IgG. FcγRIIB is expressed in two forms following alternative splicing, forming the FcγRIIB1 and FcγRIIB2 isoforms. FcγRIII is divided into two families, the transmembrane form FcγRIIIA and the glycosylphosphatidylinositol (GPI) linked FcγRIIIB. There are several forms of each of these receptors based on their glycosylation patterns.[25,26]

Given that most Fcγ receptors are transmembrane proteins, the first step of receptor activation and subsequent phagocytosis is binding of IgG containing complexes to the receptor extracellular domain. This crosslinking has been observed to induce Fcγ receptors to preferentially localize into lipid rafts.[27-30] The partitioning into lipid rafts may aid in recruiting and complexing with additional signaling proteins associated with lipid rafts.[31,32]

Molecular Mechanisms of Phagocytosis, edited by Carlos Rosales. ©2005 Eurekah.com and Springer Science+Business Media.

Table 1. Size and affinity for ligand of common Fcγ receptors

	FcγRI	FcγRIIA	FcγRIIB	FcγRIII
Molecular Mass	70 kDa	40 kDa	40 kDa	50-80 kDa
IgG Specificity	1>=3>4>2	3>1>2>4	3>=1>4>2	1=3>2=4
IgG Affinity	10^{-7}-10^{-9}M	< 10^{-7}M	< 10^{-7}M	<= 2×10^{-7}M

Receptor Activation

Initial Fcγ receptor activation takes place upon ligand binding to the extracellular domain. Since each Fcγ receptor has a structurally distinct extracellular domain, they bind to IgG with varying affinity as shown in Table 1. In addition, a link between the innate and adaptive immune system has been suggested by the ability of human Fcγ receptors to bind serum amyloid P (SAP) and C-reactive protein (CRP).[33-36] SAP and CRP bind to lectin sites on pathogens and mediate their recognition by leukocytes such as neutrophils and macrophages. Human neutrophils have been shown to bind and internalize SAP and CRP both as complexes and as an opsonin for zymosan and this is at least in part through Fcγ receptors.[33,37] In addition, binding and phagocytosis by human Fcγ receptors has been observed both in normal human leukocytes and in model systems such as transfected Cos-7 cells.[38] Therefore, in addition to mediating phagocytosis of IgG coated particles, human Fcγ receptors may play an additional role in mediating innate recognition of pathogens.

Fcγ receptors traditionally signal through an immunoreceptor tyrosine based activation motif (ITAM)[39] or through inhibitory residues found in an immunoreceptor tyrosine based inhibitory motif (ITIM).[40] The classic ITAM motif consists of two YxxL sequences separated by 7 amino acids.[41] Fcγ receptor ITAM sequences in the Fc receptor associated γ-chain abide by this structure. However, the cytoplasmic domain of FcγRIIA contains an ITAM-like domain. This ITAM-like domain contains the two YxxL motifs but a spacer sequence containing 12 amino acids instead of the usual 7.[40] ITAM tyrosine residues have been shown to be crucial for mediating the phagocytic response. When either of the ITAM tyrosines are mutated to phenylalanine phagocytosis is inhibited nearly 80%. However, if both tyrosine residues are mutated phagocytosis is abolished.[11,40,42-46] In addition, recent data suggest that the nonITAM tyrosines located upstream of the ITAM may play a role in binding of the tyrosine kinase Syk (Kim, M-K et al unpublished observations).

It has been proposed that ITAM tyrosines are phosphorylated by Src family kinases after crosslinking.[44,46-52] Src family tyrosine kinases are thought to be inactive through phosphorylation of a tyrosine at the carboxy terminal end of the protein that interacts with its own SH2 binding site. The proposed mechanism of activation of the Src family of tyrosine kinases occurs by dephosphorylation of the tyrosine (perhaps by CD45), releasing the SH2 and allowing phosphorylation of ITAM tyrosines.[53,54]

Several members of the Src family have been shown to associate with specific Fcγ receptors.[52,55-58] However, which Src kinase is responsible for phosphorylation of specific Fc receptors is more elusive. Studies have been somewhat inconclusive in elucidating which kinase is responsible for phagocytic signaling through each Fcγ receptor. An example of these observations can be found in knockout experiments where phagocytosis is not abolished in Hck, Lyn, or Fgr single knockouts.[59] In addition, in triple knockout mice, phagocytosis by macrophages and neutrophils is still partially intact suggesting other kinases may play a redundant role in phagocytosis.[50]

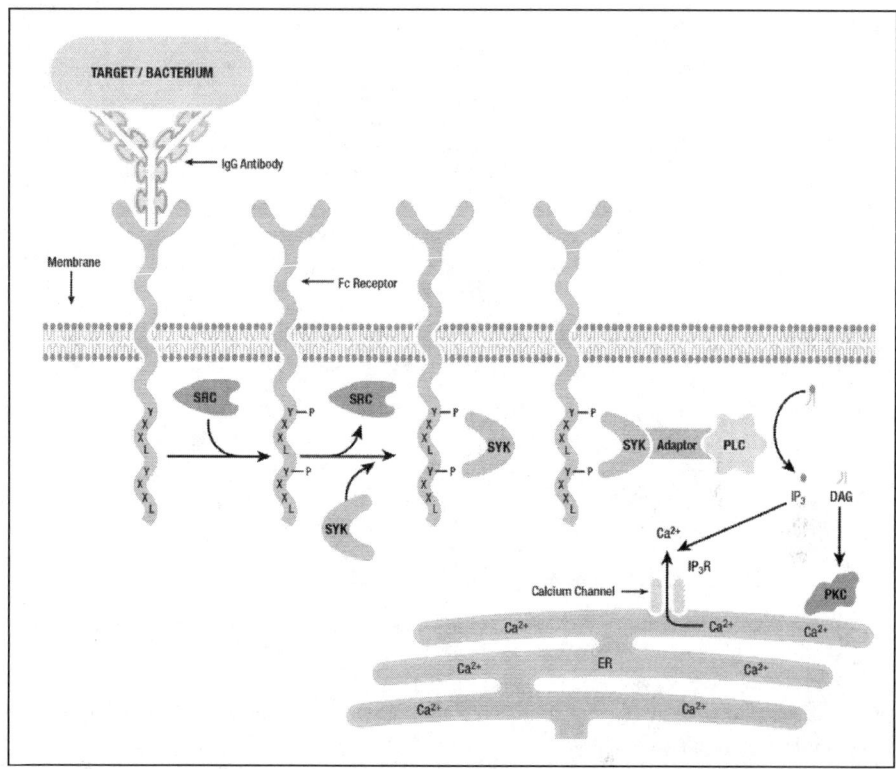

Figure 1. Binding and initial signaling after Fc receptor ligation.

Phosphorylation of ITAM tyrosines creates Src homology 2 (SH2) binding sites required for signal transduction involving other members of the tyrosine kinase family.[60-62] Most importantly, Syk tyrosine kinase, which contains two SH2 binding sites, is recruited to phosphorylated ITAM residues (Fig. 1).[61,63-66] FcγR phagocytosis is dependent upon Syk signaling, and has been observed to be enhanced upon the overexpression of Syk in model systems.[67-69] In addition, when Syk kinase expression is inhibited with antisense oligonucleotides both in vitro and in vivo, phagocytosis and inflammation is abolished.[70-72]

SH2 containing proteins are important in signaling complexes. For example, recent studies have emphasized the role of adaptor proteins in phagocytic signaling. Adaptor molecules such as SLP-76, LAT, Cbl and others have the ability to recruit SH2 containing proteins to signaling complexes, notably in lipid rafts (Fig. 1).[73-76] These adaptor proteins play a significant role in recruiting such secondary signaling molecules as phospholipase C (PLC), Grb2, Shc and others.[77,78] The ability of adaptor molecules to recruit proteins to the site of signal propagation by Fcγ receptors is important for efficiently triggering the phagocytic response and target internalization.

Phagocytic Cup Formation

After initiation of a phagocytic signal, the cell needs to prepare to internalize the bound (opsonized) particle into the cell and enclose it in a phagosome. A crucial step is the ability of a phagocytic signal to proceed from tyrosine phosphorylation to actin rearrangement for

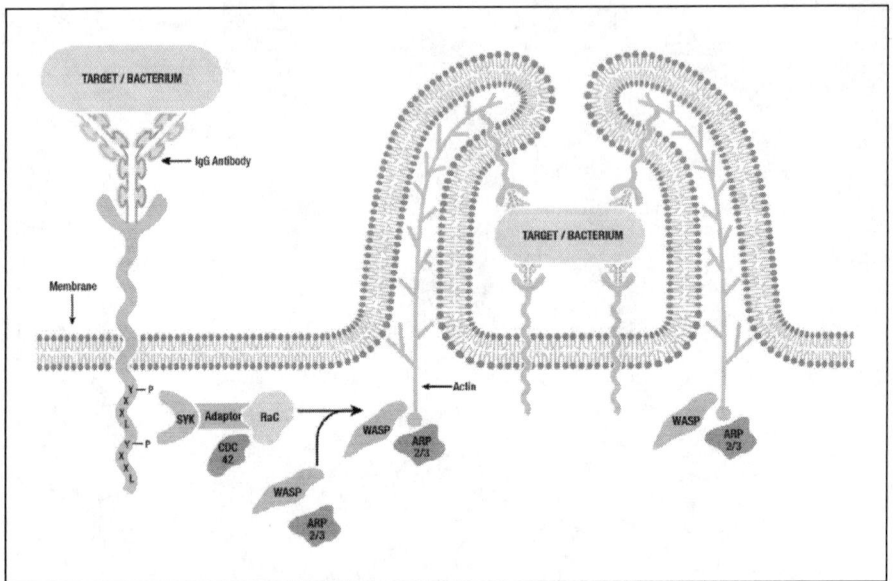

Figure 2. Membrane extension and actin polymerization leading to internalization of target bound to Fc receptor.

pseudopod formation (Fig. 2). The ability to stimulate actin polymerization occurs by an as yet incompletely understood pathway. It has been proposed that the signal sequence leading to actin polymerization includes the activation of Rho GTPases such as Rac and CDC42, PI3K, ERK, PKC and others.[79,80] The rearrangement of actin to form a phagocytic cup has been experimentally shown to be dependent on Rac and CDC42.[81-83] This Rho family of GTPases is a crucial effector of phagocytosis as inhibition of either of these GTPase enzymes results in significantly reduced phagocytosis.[84] Activation of GTPases such as Rac and CDC42 involves guanine nucleotide-exchange factors (GEF). Vav, a GEF, observed to be involved in FcγR mediated phagocytosis localizes to the site at which Rac and CDC42 are present in their inactive states and where Vav can then activate Rac and CDC42.[74,83-85] Subsequently, CDC42 accumulates at the phagocytic cup where it associates with Wiskott-Aldrich syndrome protein (WASP), which binds to both CDC42 and Rac.[86] WASP can then associate with the Arp2/3 complex, which seems to mediate the actual actin nucleation to form the phagocytic cup. Studies have shown that omission of any of these proteins in knockout experiments or in human disease results in significantly decreased phagocytosis.[86-92] The phagocytic cup is also the site at which many other molecules involved in phagocytic signaling are found. Thus the formation of a phagocytic cup at the site of binding and eventual phagocytosis is the second step in FcγR mediated phagocytosis.

Additional Signal Transduction Molecules: Lipid and Kinases

One important messenger that associates with the phagocytic cup are lipid modifying enzymes and their products. A relatively early step in phagocytic signaling is the association of PIP2 and IP$_3$ at the phagocytic cup.[93,94] Inositol derivatives are formed by the activity of two major families of enzymes, phosphoinositol kinases and phospholipases. Phosphoinositol kinases are represented by phosphoinositol 3-kinase (PI3K) which catalyzes the addition of phosphate at the 3' position of inositol.[95] This modification results in the presence of PI(3,4,5)P and PI(3,4)P. Recently, PI3K has been shown to accumulate at phagocytic cups.[94,96] In addition, PI3K is intimately involved in phagocytosis due to the observation that the addition of wortmanin, a PI3K inhibitor, abolishes phagocytosis.[94,97,98] The specific role of PI3K in phagocytosis is yet to be completely understood. PI3K may activate ERK or result in PKC activation, recruitment of PH-domain bearing proteins such as Vav and PLCγ, and PKB/Akt which play roles in actin rearrangement and nucleation as discussed earlier.[93,95,99-104]

Phospholipases (PL) are lipid modifying enzymes that result in the formation of various lipid mediators such as diacylglycerol, arachidonic acid, and IP$_3$. Of particular interest in phagocytosis are the phospholipases PLA, PLC and PLD. PLA mediates the production of arachidonic acid from phosphatidylcholine and phosphatidylethanolamine.[105,106] Recent observations suggest that inhibition of PLA inhibits phagocytosis and that phagocytosis can be reconstituted upon the addition of arachidonate.[107-109] Phospholipase C accumulates at phagocytic cups and is involved in the production of two important products, namely IP$_3$ and diacylglycerol (DAG) from the hydrolysis of PI(4,5)P.[93,110-112] DAG has been shown in various systems to activate protein kinase C (see below). IP$_3$ is responsible for triggering calcium release from the endoplasmic reticulum through ligation of IP$_3$R.[113,114] Release of calcium from intracellular stores is crucial for various signaling events such as phagolysosome fusion and PKC activation (see below). Inhibition of PLC appears to result in significant reduction of phagocytosis.[93] PLD has been observed to become activated during phagocytosis and inhibition of PLD has also been shown to inhibit phagocytosis.[115-117] PLD is responsible for producing phosphatidic acid and choline of which phosphatidic acid can be converted to DAG to activate PKC while choline is capable of activating PLC and PLA, an example of redundancy in signaling pathways leading to phagocytosis.[106]

Calcium Signaling

The role of calcium during phagocytosis is controversial. Phagocytosis has been shown to be both calcium dependent and calcium independent depending on the cell type studied and the specific FcγR mediating the phagocytic signal. For example, it has been proposed that phagocytosis mediated by γ-chain associated receptors does not require calcium.[118] In addition, two different groups have suggested both calcium dependent and calcium independent phagocytosis by human neutrophils.[119,120] Results have also reported calcium independence of phagocytosis in macrophages and monocytes.[121,122] Calcium transients observed using high-speed imaging systems after phagocytosis in both neutrophils and in model systems such as Chinese hamster ovary cells suggest that calcium plays a significant role in phagocytosis.[123,124]

Calcium is involved in many processes that are required for phagocytosis. Intracellular calcium concentrations have been observed to be of greatest intensity near phagosomes and phagocytic cups.[125] Calcium release from phagosomes into the cytoplasm after FcγR mediated phagocytosis has been observed.[126] These data are consistent with a recent observation suggesting that the endoplasmic reticulum (ER) plays a role in phagosome formation.[127] If the ER plays a part in phagosome formation then calcium released from the phagosome may be analogous to calcium released from ER (Fig. 3). Additionally, calcium waves have been observed to encircle phagosomes after FcγR mediated phagocytosis in both neutrophils and FcγRIIA expressing CHO cells.[123,124]

Calcium near the phagocytic cup and phagosomes may play a crucial role in both actin cytoskeleton depolymerization and reformation. Calcium has been shown to activate gelsolin, which in an active state, binds to the barbed ends of actin filaments preventing polymerization and promoting depolymerization by severing actin filaments.[128,129] Calcium transients surrounding the phagosome after phagocytosis may play a role in mediating vesicle fusion events during phagosome-lysosome fusion (Fig. 3). Recent observations suggest that mutation of a vital LTL motif in the cytoplasmic domain of FcγRIIA inhibits phagolysosome fusion. This

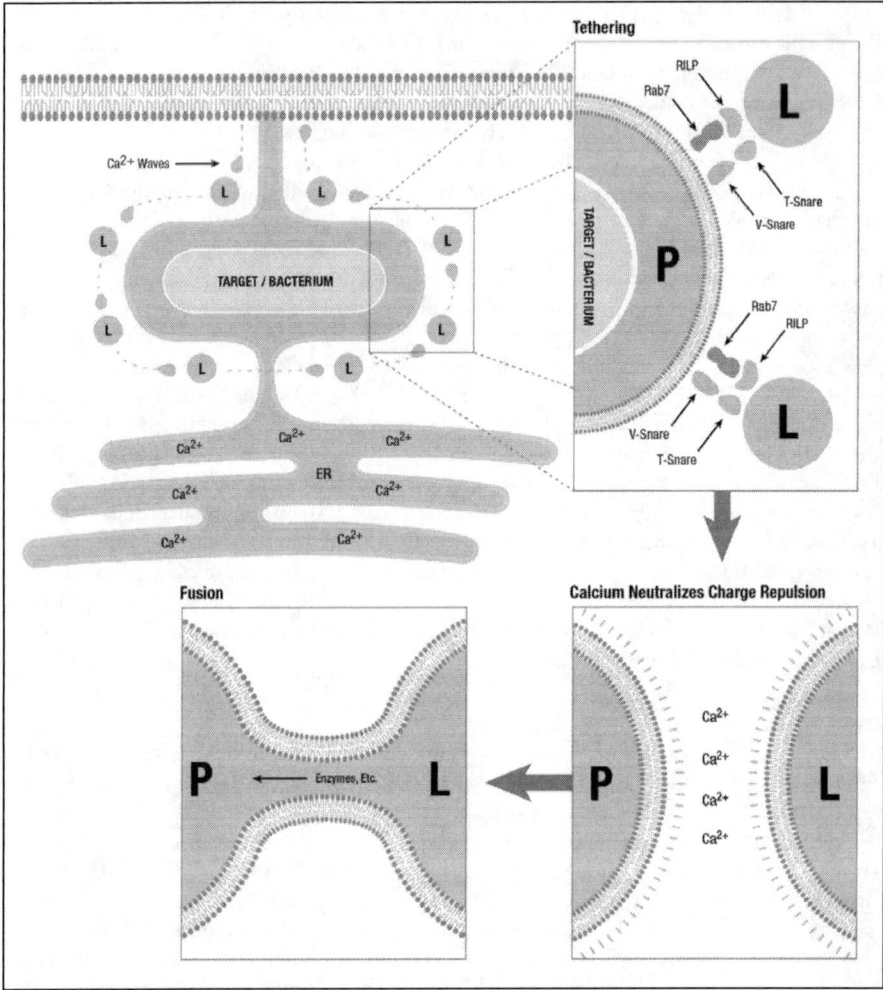

Figure 3. Model of phagolysosome fusion after Fc receptor mediated phagocytosis.

lesion is due to the inability of the receptor to direct calcium waves to the phagosome suggesting that calcium routing is a crucial step in FcγR mediated phagocytosis and eventual phagosome-lysosome fusion.[124] This observation shows the intricate regulation of FcR mediated processes and how seemingly minimal changes can affect the efficiency of phagocytosis.

Phagosome Formation and Fusion

Once a target is bound and the activation signals have provided the stimulation necessary to mediate phagocytosis, the phagosome must then be taken into the cytosol and detached from the plasma membrane. As stated earlier actin rearrangement is crucial to this process. Upon remodeling of the actin cytoskeleton by such molecules as gelsolin, the phagosome is taken into the cytosol driven by other actin binding molecules. One such molecule is myosin, which is a downstream effector from PI3K and PKC.[102,103,130-132] PI3K and PKC activate ERK which in turn activates myosin light chain kinase which phosphorylates myosin.[133,134] Myosin then binds actin and mediates a process needed to stimulate phagosome entry into the cell.[96,135-138] Various myosins have been shown to associate with phagocytic cups and phagosomes.[139-142] Therefore, several different mysoins may play roles at different steps along the phagocytic process. Once the phagosome is pulled into the cell, the phagosomal vacuole needs to close and pinch off from the plasma membrane.[142] This activity has been linked to several other molecules including PI3K and a component of the actin cytoskeleton, dynamin.[143,144] Inhibition of PI3K has been observed to inhibit macropinocytosis and phagocytosis using inhibitors such as wortmannin.[144] In this study, wortmanin does not inhibit pseudopod extension but does inhibit phagosome closure and fusion from the plasma membrane.[144] A role for dynamin has also been proposed. Dynamin has been suggested to wrap itself around the linking region of membrane connecting the endocytic vacuole and the plasma membrane.[145] Dynamin then contracts and eventually pinches off the vacuole to form a stand-alone phagosome. However, recent observations suggest that other dynamin independent mechanisms play a role in phagocytosis and eventual phagosome closure.[146] A separate family of proteins mediates the membrane fusion event required for lipid bilayer detachment. Soluble NSF attachment receptor (SNARE) proteins compose a family of proteins involved in mediating membrane fusion events. An interaction between vSNARE and tSNARE on vesicles and target membranes respectively is dependent on the complexing of these two proteins.[147-150] When vSNARE and tSNARE interact, the lipid bilayers come into close contact and are able to fuse forming an intracellular vesicle, the phagosome, and closing the plasma membrane.

Once a phagosome is separated from the plasma membrane, it can undergo fusion events leading to the eventual destruction of the phagocytosed particle. The "maturation" of the phagosome occurs through interaction with the endocytic compartment. This has been visualized by the transient recruitment of various phosphoinositol products such as PI(4,5)P, PI(3,4,5)P, and PI(3)P at various times post phagosome formation.[151] As proposed by several investigators, phagosomes acquire markers of early and late endosomes and lysosomes in an orchestrated manner (Fig. 3).[152-154] As such, Rab5 has been observed to accumulate on early phagosomes and regulates fusion between phagosomes and early endosomes.[153,155-157] Rab5 recruits another endosome molecule, early endosome antigen 1 (EEA1) which binds Rab5 through its two Rab5 binding domains and contains another PI3P binding domain termed a FYVE domain.[158,159] Upon Rab5 dissociation, Rab7 is recruited.[152] Rab7 is a late endosome marker and has been associated with acquisition of Rab7-interacting lysosomal protein (RILP) which has been implicated in fusion of endosomes with lysosomes.[160,161] These intermediate steps suggest that phagosome lysosome/endosome interactions take place transiently and may act through a "kiss and run" mechanism.

Data supporting a "kiss and run" hypothesis can be found in various studies looking at the composition of maturing phagosomes. Elegant work by Desjardins et al has observed microdomains on the phagosome membrane.[162,163] These microdomains contain various proteins and lipids normally associated with lipid rafts and therefore may have a significant function on the phagosome membrane.

Fc Receptor Interacting Proteins

On the surface of leukocytes, there are several other proteins that have been suggested to associate with Fc receptors. These molecules have often been referred to as "accessory" due to the ability of Fc receptors to function in the absence of these proteins. However, it has been observed that the intermolecular interactions between Fc receptors and other surface receptors play an important role in mediating optimal Fc receptor mediated activation. A good example of Fc receptor interactions are the family of Fcγ receptors that interact with CR3, a member of the β2 integrin family. It was initially observed that receptors for IgG and receptors for complement interact during recognition and phagocytosis of IgG and complement opsonized targets.[164] Since these initial observations, much has been done to understand the mechanics behind these interactions. Some of the most understood interactions are between the glycosylphosphatidylinositol-linked Fcγ receptor, FcγRIIIB (CD64b) and complement receptor type 3 (CR3, Mac-1, αmβ2, CD11bCD18). FcγRIIIB was first observed to interact with CR3 in cocapping experiments.[165] This interaction was further confirmed by fluorescence resonance energy transfer experiments in both human neutrophils and transfectant model systems, and by observing that soluble FcγRIIIB binds to CR3 via flow cytometry.[166-169] In addition, in a model transfectant system it was observed that FcγRIIIB, which does not contain a cytoplasmic domain, can mediate phagocytosis of IgG-coated targets when cotransfected with CR3.[169]

Other Fcγ receptors have also been observed to interact with complement receptors. Specifically, the transmembrane receptor FcγRIIA is in physical proximity to CR3 on neutrophils and in model systems of transfectant receptors.[165,170] CR3 also plays a role in enhancing FcγRIIA and FcγRIIIB mediated phagocytosis and oxidative burst.[171,172] Interestingly, patients with leukocyte adhesion deficiency (e.g., a deficiency in β2 integrins) show a substantial reduction in phagocytosis and calcium signaling.[171,173,174]

Inhibitory/Regulatory Fc Receptors

Activation of Fcγ receptors is important for mediating a phagocytic response. However, inhibitory signals can be mediated by certain Fcγ receptors that aid in regulating the phagocytic response. Inhibitory receptors such as FcγRIIB signal through an immunoreceptor tyrosine-based inhibitory motif (ITIM) and has been shown to recruit Src homology 2 domain-containing inositol 5'-phosphatase (SHIP) and SHP-2.[175-178] Observations made using catalytically inactive SHIP expressing cells have shown that phagocytosis is enhanced in the absence of SHIP in both Fc receptor and complement receptor mediated phagocytosis.[178] Tyrosine phosphatases such as SHP-1 and SHP-2 dephosphorylate tyrosine residues of adaptor molecule(s) such as Cbl, SLP-76 and CRKL and therefore inhibit further phagocytic signaling by inhibiting Rac activation.[179,180] SHIP removes the 5'phosphate from PIP3 therefore inhibiting PKC and other signaling molecules, thus inhibiting phagocytosis. In addition, SHIP and SHP-1 have both been observed to associate with the activating receptors FcγRI and FcγRIIA upon ligation.[181,182] That phosphatases such as these are capable of interacting with activating receptors suggest a dichotomy of signaling including activation and self-regulation. Further work is undergoing to elucidate the mechanisms of possible self-regulation and the role of phosphatases in Fc receptor-mediated phagocytosis.

In summary, Fc receptor mediated phagocytosis is known to involve several steps. However, these studies are still in the early stages. Research in all of the areas above should help to elucidate how the phagocytic response is regulated.

References

1. Hulett MD, Hogarth PM. Molecular basis of Fc receptor function. Adv Immunol 1994; 57:1-127.
2. Unkeless JC, Scigliano E, Freedman VH. Structure and function of human and murine receptors for IgG. Annu Rev Immunol 1988; 6:251-281.
3. Ravetch JV, Kinet JP. Fc receptors. Annu Rev Immunol 1991; 9:457-492.
4. Fridman WH. Fc receptors and immunoglobulin binding factors. FASEB J 1991; 5(12):2684-2690.
5. Amigorena S, Salamero J, Davoust J et al. Tyrosine-containing motif that transduces cell activation signals also determines internalization and antigen presentation via type III receptors for IgG. Nature 1992; 358(6384):337-341.
6. Gosselin EJ, Wardwell K, Gosselin DR et al. Enhanced antigen presentation using human Fc gamma receptor (monocyte/macrophage)-specific immunogens. J Immunol 1992; 149(11):3477-3481.
7. Mao SY, Varin-Blank N, Edidin M et al. Immobilization and internalization of mutated IgE receptors in transfected cells. J Immunol 1991; 146(3):958-966.

8. Silvain C, Patry C, Launay P et al. Altered expression of monocyte IgA Fc receptors is associated with defective endocytosis in patients with alcoholic cirrhosis. Potential role for IFN-gamma. J Immunol 1995; 155(3):1606-1618.

9. Anderson CL, Shen L, Eicher DM et al. Phagocytosis mediated by three distinct Fc gamma receptor classes on human leukocytes. J Exp Med 1990; 171(4):1333-1345.

10. Yeaman GR, Kerr MA. Opsonization of yeast by human serum IgA anti-mannan antibodies and phagocytosis by human polymorphonuclear leucocytes. Clin Exp Immunol 1987; 68(1):200-208.

11. Daeron M, Malbec O, Bonnerot C et al. Tyrosine-containing activation motif-dependent phagocytosis in mast cells. J Immunol 1994; 152(2):783-792.

12. Tuijnman WB, Capel PJ, van de Winkel JG. Human low-affinity IgG receptor Fc gamma RIIa (CD32) introduced into mouse fibroblasts mediates phagocytosis of sensitized erythrocytes. Blood 1992; 79(7):1651-1656.

13. Daeron M, Latour S, Huckel C et al. Murine Fc gamma RII and III in mast cell activation. Immunobiology 1992; 185(2-4):159-174.

14. Dombrowicz D, Flamand V, Brigman KK et al. Abolition of anaphylaxis by targeted disruption of the high affinity immunoglobulin E receptor alpha chain gene. Cell 1993; 75(5):969-976.

15. Takai T, Li M, Sylvestre D et al. FcR gamma chain deletion results in pleiotrophic effector cell defects. Cell 1994; 76(3):519-529.

16. Vaz NM, Prouvost-Danon A. Behaviour of mouse mast cells during anaphylaxis in vitro. Prog Allergy 1969; 13:111-173.

17. Alber G, Kent UM, Metzger H. Functional comparison of Fc epsilon RI, Fc gamma RII, and Fc gamma RIII in mast cells. J Immunol 1992; 149(7):2428-2436.

18. Hazenbos WL, Gessner JE, Hofhuis FM et al. Impaired IgG-dependent anaphylaxis and Arthus reaction in Fc gamma RIII (CD16) deficient mice. Immunity 1996; 5(2):181-188.

19. Vivier E, Rochet N, Ackerly M et al. Signaling function of reconstituted CD16: Zeta: Gamma receptor complex isoforms. Int Immunol 1992; 4(11):1313-1323.

20. Patry C, Herbelin A, Lehuen A et al. Fc alpha receptors mediate release of tumour necrosis factor-alpha and interleukin-6 by human monocytes following receptor aggregation. Immunology 1995; 86(1):1-5.

21. Fanger MW, Goldstine SN, Shen L. The properties and role of receptors for IgA on human leukocytes. Ann NY Acad Sci 1983; 409:552-563.

22. Zhou MJ, Lublin DM, Link DC et al. Distinct tyrosine kinase activation and Triton X-100 insolubility upon Fc gamma RII or Fc gamma RIIIB ligation in human polymorphonuclear leukocytes. Implications for immune complex activation of the respiratory burst. J Biol Chem 1995; 270(22):13553-13560.

23. Tomiyama Y, Kunicki TJ, Zipf TF et al. Response of human platelets to activating monoclonal antibodies: Importance of Fc gamma RII (CD32) phenotype and level of expression. Blood 1992; 80(9):2261-2268.

24. Ravetch JV, Bolland S. IgG Fc receptors. Annu Rev Immunol 2001; 19:275-290.

25. Huizinga TW, Kleijer M, Tetteroo PA et al. Biallelic neutrophil Na-antigen system is associated with a polymorphism on the phospho-inositol-linked Fc gamma receptor III (CD16). Blood 1990; 75(1):213-217.

26. Ory PA, Clark MR, Talhouk AS et al. Transfected NA1 and NA2 forms of human neutrophil Fc receptor III exhibit antigenic and structural heterogeneity. Blood 1991; 77(12):2682-2687.

27. Kono H, Suzuki T, Yamamoto K et al. Spatial raft coalescence represents an initial step in Fc gamma R signaling. J Immunol 2002; 169(1):193-203.

28. Suzuki T, Kono H, Hirose N et al. Differential involvement of Src family kinases in Fc gamma receptor-mediated phagocytosis. J Immunol 2000; 165(1):473-482.

29. Katsumata O, Hara-Yokoyama M, Sautes-Fridman C et al. Association of FcgammaRII with low-density detergent-resistant membranes is important for cross-linking-dependent initiation of the tyrosine phosphorylation pathway and superoxide generation. J Immunol 2001; 167(10):5814-5823.

30. Barabe F, Rollet-Labelle E, Gilbert C et al. Early events in the activation of Fc gamma RIIA in human neutrophils: Stimulated insolubilization, translocation to detergent-resistant domains, and degradation of Fc gamma RIIA. J Immunol 2002; 168(8):4042-4049.

31. Katagiri YU, Kiyokawa N, Fujimoto J. A role for lipid rafts in immune cell signaling. Microbiol Immunol 2001; 45(1):1-8.
32. Liang X, Nazarian A, Erdjument-Bromage H et al. Heterogeneous fatty acylation of Src family kinases with polyunsaturated fatty acids regulates raft localization and signal transduction. J Biol Chem 2001; 276(33):30987-30994.
33. Mortensen RF, Duszkiewicz JA. Mediation of CRP-dependent phagocytosis through mouse macrophage Fc-receptors. J Immunol 1977; 119(5):1611-1616.
34. Mold C, Gresham HD, Du Clos TW. Serum amyloid P component and C-reactive protein mediate phagocytosis through murine Fc gamma Rs. J Immunol 2001; 166(2):1200-1205.
35. Chi M, Tridandapani S, Zhong W et al. C-reactive protein induces signaling through Fc gamma RIIa on HL-60 granulocytes. J Immunol 2002; 168(3):1413-1418.
36. Stein MP, Edberg JC, Kimberly RP et al. C-reactive protein binding to FcgammaRIIa on human monocytes and neutrophils is allele-specific. J Clin Invest 2000; 105(3):369-376.
37. Bharadwaj D, Mold C, Markham E et al. Serum amyloid P component binds to Fc gamma receptors and opsonizes particles for phagocytosis. J Immunol 2001; 166(11):6735-6741.
38. Marnell LL, Mold C, Volzer MA et al. C-reactive protein binds to Fc gamma RI in transfected COS cells. J Immunol 1995; 155(4):2185-2193.
39. Cambier JC. New nomenclature for the Reth motif (or ARH1/TAM/ARAM/YXXL). Immunol Today 1995; 16(2):110.
40. Van den Herik-Oudijk IE, Capel PJ, van der Bruggen T et al. Identification of signaling motifs within human Fc gamma RIIa and Fc gamma RIIb isoforms. Blood 1995; 85(8):2202-2211.
41. Reth M. Antigen receptor tail clue. Nature 1989; 338(6214):383-384.
42. Mitchell MA, Huang MM, Chien P et al. Substitutions and deletions in the cytoplasmic domain of the phagocytic receptor Fc gamma RIIA: Effect on receptor tyrosine phosphorylation and phagocytosis. Blood 1994; 84(6):1753-1759.
43. Indik ZK, Mitchell MA, Chien P et al. Structural requirements for phagocytosis by the human Fc receptor Fc gamma RIIA. Trans Assoc Am Physicians 1993; 106:77-85.
44. Strzelecka A, Kwiatkowska K, Sobota A. Tyrosine phosphorylation and Fcgamma receptor-mediated phagocytosis. FEBS Lett 1997; 400(1):11-14.
45. Park JG, Murray RK, Chien P et al. Conserved cytoplasmic tyrosine residues of the gamma subunit are required for a phagocytic signal mediated by Fc gamma RIIIA. J Clin Invest 1993; 92(4):2073-2079.
46. Darby C, Geahlen RL, Schreiber AD. Stimulation of macrophage Fc gamma RIIIA activates the receptor-associated protein tyrosine kinase Syk and induces phosphorylation of multiple proteins including p95Vav and p62/GAP-associated protein. J Immunol 1994; 152(11):5429-5437.
47. Ibarrola I, Vosseberg PJ, Homburg CH et al. Influence of tyrosine phosphorylation on protein interaction with FcgammaRIIa. Biochim Biophys Acta 1997; 1357(3):348-358.
48. Isakov N. Immunoreceptor tyrosine-based activation motif (ITAM), a unique module linking antigen and Fc receptors to their signaling cascades. J Leukoc Biol 1997; 61(1):6-16.
49. Flaswinkel H, Barner M, Reth M. The tyrosine activation motif as a target of protein tyrosine kinases and SH2 domains. Semin Immunol 1995; 7(1):21-27.
50. Fitzer-Attas CJ, Lowry M, Crowley MT et al. Fcgamma receptor-mediated phagocytosis in macrophages lacking the Src family tyrosine kinases Hck, Fgr, and Lyn. J Exp Med 2000; 191(4):669-682.
51. Kwiatkowska K, Frey J, Sobota A. Phosphorylation of FcgammaRIIA is required for the receptor-induced actin rearrangement and capping: The role of membrane rafts. J Cell Sci 2003; 116(Pt 3):537-550.
52. Korade-Mirnics Z, Corey SJ. Src kinase-mediated signaling in leukocytes. J Leukoc Biol 2000; 68(5):603-613.
53. Erpel T, Courtneidge SA. Src family protein tyrosine kinases and cellular signal transduction pathways. Curr Opin Cell Biol 1995; 7(2):176-182.
54. Adamczewski M, Numerof RP, Koretzky GA et al. Regulation by CD45 of the tyrosine phosphorylation of high affinity IgE receptor beta- and gamma-chains. J Immunol 1995; 154(7):3047-3055.
55. Ghazizadeh S, Bolen JB, Fleit HB. Physical and functional association of Src-related protein tyrosine kinases with Fc gamma RII in monocytic THP-1 cells. J Biol Chem 1994; 269(12):8878-8884.

56. Durden DL, Kim HM, Calore B et al. The Fc gamma RI receptor signals through the activation of hck and MAP kinase. J Immunol 1995; 154(8):4039-4047.

57. Hamada F, Aoki M, Akiyama T et al. Association of immunoglobulin G Fc receptor II with Src-like protein-tyrosine kinase Fgr in neutrophils. Proc Natl Acad Sci USA 1993; 90(13):6305-6309.

58. Pignata C, Prasad KV, Robertson MJ et al. Fc gamma RIIIA-mediated signaling involves src-family lck in human natural killer cells. J Immunol 1993; 151(12):6794-6800.

59. Hunter S, Huang MM, Indik ZK et al. Fc gamma RIIA-mediated phagocytosis and receptor phosphorylation in cells deficient in the protein tyrosine kinase Src. Exp Hematol 1993; 21(11):1492-1497.

60. Johnson SA, Pleiman CM, Pao L et al. Phosphorylated immunoreceptor signaling motifs (ITAMs) exhibit unique abilities to bind and activate Lyn and Syk tyrosine kinases. J Immunol 1995; 155(10):4596-4603.

61. Ghazizadeh S, Bolen JB, Fleit HB. Tyrosine phosphorylation and association of Syk with Fc gamma RII in monocytic THP-1 cells. Biochem J 1995; 305(Pt 2):669-674.

62. Kimura T, Sakamoto H, Appella E et al. Conformational changes induced in the protein tyrosine kinase p72syk by tyrosine phosphorylation or by binding of phosphorylated immunoreceptor tyrosine-based activation motif peptides. Mol Cell Biol 1996; 16(4):1471-1478.

63. Turner M, Schweighoffer E, Colucci F et al. Tyrosine kinase SYK: Essential functions for immunoreceptor signalling. Immunol Today 2000; 21(3):148-154.

64. Cooney DS, Phee H, Jacob A et al. Signal transduction by human-restricted Fc gamma RIIa involves three distinct cytoplasmic kinase families leading to phagocytosis. J Immunol 2001; 167(2):844-854.

65. Chacko GW, Duchemin AM, Coggeshall KM et al. Clustering of the platelet Fc gamma receptor induces noncovalent association with the tyrosine kinase p72syk. J Biol Chem 1994; 269(51):32435-32440.

66. Jouvin MH, Adamczewski M, Numerof R et al. Differential control of the tyrosine kinases Lyn and Syk by the two signaling chains of the high affinity immunoglobulin E receptor. J Biol Chem 1994; 269(8):5918-5925.

67. Greenberg S, Chang P, Silverstein SC. Tyrosine phosphorylation of the gamma subunit of Fc gamma receptors, p72syk, and paxillin during Fc receptor-mediated phagocytosis in macrophages. J Biol Chem 1994; 269(5):3897-3902.

68. Greenberg S, Chang P, Wang DC et al. Clustered syk tyrosine kinase domains trigger phagocytosis. Proc Natl Acad Sci USA 1996; 93(3):1103-1107.

69. Indik ZK, Park JG, Pan XQ et al. Induction of phagocytosis by a protein tyrosine kinase. Blood 1995; 85(5):1175-1180.

70. Matsuda M, Park JG, Wang DC et al. Abrogation of the Fc gamma receptor IIA-mediated phagocytic signal by stem-loop Syk antisense oligonucleotides. Mol Biol Cell 1996; 7(7):1095-1106.

71. Stenton GR, Kim MK, Nohara O et al. Aerosolized Syk antisense suppresses Syk expression, mediator release from macrophages, and pulmonary inflammation. J Immunol 2000; 164(7):3790-3797.

72. Stenton GR, Ulanova M, Dery RE et al. Inhibition of allergic inflammation in the airways using aerosolized antisense to Syk kinase. J Immunol 2002; 169(2):1028-1036.

73. Tridandapani S, Lyden TW, Smith JL et al. The adapter protein LAT enhances fcgamma receptor-mediated signal transduction in myeloid cells. J Biol Chem 2000; 275(27):20480-20487.

74. Coppolino MG, Krause M, Hagendorff P et al. Evidence for a molecular complex consisting of Fyb/SLAP, SLP-76, Nck, VASP and WASP that links the actin cytoskeleton to Fcgamma receptor signalling during phagocytosis. J Cell Sci 2001; 114(Pt 23):4307-4318.

75. Bonilla FA, Fujita RM, Pivniouk VI et al. Adapter proteins SLP-76 and BLNK both are expressed by murine macrophages and are linked to signaling via Fcgamma receptors I and II/III. Proc Natl Acad Sci USA 2000; 97(4):1725-1730.

76. Chu J, Liu Y, Koretzky GA et al. SLP-76-Cbl-Grb2-Shc interactions in FcgammaRI signaling. Blood 1998; 92(5):1697-1706.

77. Saxton TM, van Oostveen I, Bowtell D et al. B cell antigen receptor cross-linking induces phosphorylation of the p21ras oncoprotein activators SHC and mSOS1 as well as assembly of complexes containing SHC, GRB-2, mSOS1, and a 145-kDa tyrosine-phosphorylated protein. J Immunol 1994; 153(2):623-636.

78. Kumar G, Wang S, Gupta S et al. The membrane immunoglobulin receptor utilizes a Shc/Grb2/ hSOS complex for activation of the mitogen-activated protein kinase cascade in a B-cell line. Biochem J 1995; 307(Pt 1):215-223.

79. Ridley AJ. Rho proteins, PI 3-kinases, and monocyte/macrophage motility. FEBS Lett 2001; 498(2-3):168-171.

80. Garcia-Garcia E, Rosales C. Signal transduction during Fc receptor-mediated phagocytosis. J Leukoc Biol 2002; 72(6):1092-1108.

81. Chimini G, Chavrier P. Function of Rho family proteins in actin dynamics during phagocytosis and engulfment. Nat Cell Biol 2000; 2(10):E191-196.

82. Hackam DJ, Rotstein OD, Schreiber A et al. Rho is required for the initiation of calcium signaling and phagocytosis by Fcgamma receptors in macrophages. J Exp Med 1997; 186(6):955-966.

83. Caron E, Hall A. Identification of two distinct mechanisms of phagocytosis controlled by different Rho GTPases. Science 1998; 282(5394):1717-1721.

84. Cox D, Chang P, Zhang Q et al. Requirements for both Rac1 and Cdc42 in membrane ruffling and phagocytosis in leukocytes. J Exp Med 1997; 186(9):1487-1494.

85. Patel JC, Hall A, Caron E. Vav regulates activation of Rac but not Cdc42 during FcgammaR-mediated phagocytosis. Mol Biol Cell 2002; 13(4):1215-1226.

86. Castellano F, Montcourrier P, Guillemot JC et al. Inducible recruitment of Cdc42 or WASP to a cell-surface receptor triggers actin polymerization and filopodium formation. Curr Biol 1999; 9(7):351-360.

87. Aspenstrom P, Lindberg U, Hall A. Two GTPases, Cdc42 and Rac, bind directly to a protein implicated in the immunodeficiency disorder Wiskott-Aldrich syndrome. Curr Biol 1996; 6(1):70-75.

88. Machesky LM, Insall RH. Scar1 and the related Wiskott-Aldrich syndrome protein, WASP, regulate the actin cytoskeleton through the Arp2/3 complex. Curr Biol 1998; 8(25):1347-1356.

89. Marchand JB, Kaiser DA, Pollard TD et al. Interaction of WASP/Scar proteins with actin and vertebrate Arp2/3 complex. Nat Cell Biol 2001; 3(1):76-82.

90. Lorenzi R, Brickell PM, Katz DR et al. Wiskott-Aldrich syndrome protein is necessary for efficient IgG-mediated phagocytosis. Blood 2000; 95(9):2943-2946.

91. Welch MD, Iwamatsu A, Mitchison TJ. Actin polymerization is induced by Arp2/3 protein complex at the surface of Listeria monocytogenes. Nature 1997; 385(6613):265-269.

92. Machesky LM, Mullins RD, Higgs HN et al. Scar, a WASp-related protein, activates nucleation of actin filaments by the Arp2/3 complex. Proc Natl Acad Sci USA 1999; 96(7):3739-3744.

93. Botelho RJ, Teruel M, Dierckman R et al. Localized biphasic changes in phosphatidylinositol-4,5-bisphosphate at sites of phagocytosis. J Cell Biol 2000; 151(7):1353-1368.

94. Marshall JG, Booth JW, Stambolic V et al. Restricted accumulation of phosphatidylinositol 3-kinase products in a plasmalemmal subdomain during Fc gamma receptor-mediated phagocytosis. J Cell Biol 2001; 153(7):1369-1380.

95. Toker A, Cantley LC. Signalling through the lipid products of phosphoinositide-3-OH kinase. Nature 1997; 387(6634):673-676.

96. Araki N, Hatae T, Furukawa A et al. Phosphoinositide-3-kinase-independent contractile activities associated with Fcgamma-receptor-mediated phagocytosis and macropinocytosis in macrophages. J Cell Sci 2003; 116(Pt 2):247-257.

97. Indik ZK, Park JG, Hunter S et al. The molecular dissection of Fc gamma receptor mediated phagocytosis. Blood 1995; 86(12):4389-4399.

98. Cox D, Tseng CC, Bjekic G et al. A requirement for phosphatidylinositol 3-kinase in pseudopod extension. J Biol Chem 1999; 274(3):1240-1247.

99. Singh SS, Chauhan A, Brockerhoff H et al. Activation of protein kinase C by phosphatidylinositol 3,4,5-trisphosphate. Biochem Biophys Res Commun 1993; 195(1):104-112.

100. Toker A, Meyer M, Reddy KK et al. Activation of protein kinase C family members by the novel polyphosphoinositides PtdIns-3,4-P2 and PtdIns-3,4,5-P3. J Biol Chem 1994; 269(51):32358-32367.

101. Wymann MP, Sozzani S, Altruda F et al. Lipids on the move: Phosphoinositide 3-kinases in leukocyte function. Immunol Today 2000; 21(6):260-264.

102. Cox D, Berg JS, Cammer M et al. Myosin X is a downstream effector of PI(3)K during phagocytosis. Nat Cell Biol 2002; 4(7):469-477.

103. Garcia-Garcia E, Rosales R, Rosales C. Phosphatidylinositol 3-kinase and extracellular signal-regulated kinase are recruited for Fc receptor-mediated phagocytosis during monocyte-to-macrophage differentiation. J Leukoc Biol 2002; 72(1):107-114.

104. Coxon PY, Rane MJ, Powell DW et al. Differential mitogen-activated protein kinase stimulation by Fc gamma receptor IIa and Fc gamma receptor IIIb determines the activation phenotype of human neutrophils. J Immunol 2000;164(12):6530-6537.

105. Gijon MA, Leslie CC. Regulation of arachidonic acid release and cytosolic phospholipase A2 activation. J Leukoc Biol 1999; 65(3):330-336.

106. Lennartz MR. Phospholipases and phagocytosis: The role of phospholipid-derived second messengers in phagocytosis. Int J Biochem Cell Biol 1999; 31(3-4):415-430.

107. Lennartz MR, Brown EJ. Arachidonic acid is essential for IgG Fc receptor-mediated phagocytosis by human monocytes. J Immunol 1991; 147(2):621-626.

108. Lennartz MR, Lefkowith JB, Bromley FA et al. Immunoglobulin G-mediated phagocytosis activates a calcium-independent, phosphatidylethanolamine-specific phospholipase. J Leukoc Biol 1993; 54(5):389-398.

109. Lennartz MR, Yuen AF, Masi SM et al. Phospholipase A2 inhibition results in sequestration of plasma membrane into electronlucent vesicles during IgG-mediated phagocytosis. J Cell Sci 1997; 110(Pt 17):2041-2052.

110. Azzoni L, Kamoun M, Salcedo TW et al. Stimulation of Fc gamma RIIIA results in phospholipase C-gamma 1 tyrosine phosphorylation and p56lck activation. J Exp Med 1992; 176(6):1745-1750.

111. Liao F, Shin HS, Rhee SG. Tyrosine phosphorylation of phospholipase C-gamma 1 induced by cross-linking of the high-affinity or low-affinity Fc receptor for IgG in U937 cells. Proc Natl Acad Sci USA 1992; 89(8):3659-3663.

112. Shen Z, Lin CT, Unkeless JC. Correlations among tyrosine phosphorylation of Shc, p72syk, PLC-gamma 1, and [Ca2+]i flux in Fc gamma RIIA signaling. J Immunol 1994; 152(6):3017-3023.

113. Mignery GA, Sudhof TC, Takei K et al. Putative receptor for inositol 1,4,5-trisphosphate similar to ryanodine receptor. Nature 1989; 342(6246):192-195.

114. Mignery GA, Sudhof TC. The ligand binding site and transduction mechanism in the inositol-1,4,5-triphosphate receptor. EMBO J 1990; 9(12):3893-3898.

115. Suchard SJ, Mansfield PJ, Boxer LA et al. Mitogen-activated protein kinase activation during IgG-dependent phagocytosis in human neutrophils: Inhibition by ceramide. J Immunol 1997; 158(10):4961-4967.

116. Melendez A, Floto RA, Gillooly DJ et al. FcgammaRI coupling to phospholipase D initiates sphingosine kinase-mediated calcium mobilization and vesicular trafficking. J Biol Chem 1998; 273(16):9393-9402.

117. Melendez A, Floto RA, Cameron AJ et al. A molecular switch changes the signalling pathway used by the Fc gamma RI antibody receptor to mobilise calcium. Curr Biol 1998; 8(4):210-221.

118. Edberg JC, Lin CT, Lau D et al. The Ca2+ dependence of human Fc gamma receptor-initiated phagocytosis. J Biol Chem 1995; 270(38):22301-22307.

119. Kobayashi K, Takahashi K, Nagasawa S. The role of tyrosine phosphorylation and Ca2+ accumulation in Fc gamma-receptor-mediated phagocytosis of human neutrophils. J Biochem (Tokyo)1995; 117(6):1156-1161.

120. Della Bianca V, Grzeskowiak M, Rossi F. Studies on molecular regulation of phagocytosis and activation of the NADPH oxidase in neutrophils. IgG- and C3b-mediated ingestion and associated respiratory burst independent of phospholipid turnover and Ca2+ transients. J Immunol 1990; 144(4):1411-1417.

121. McNeil PL, Swanson JA, Wright SD et al. Fc-receptor-mediated phagocytosis occurs in macrophages without an increase in average [Ca++]i. J Cell Biol 1986; 102(5):1586-1592.

122. Zimmerli S, Majeed M, Gustavsson M et al. Phagosome-lysosome fusion is a calcium-independent event in macrophages. J Cell Biol 1996; 132(1-2):49-61.

123. Kindzelskii AL, Petty HR. Intracellular calcium waves accompany neutrophil polarization, formylmethionylleucylphenylalanine stimulation, and phagocytosis: A high speed microscopy study. J Immunol 2003; 170(1):64-72.

124. Worth RG, Kim MK, Kindzelskii AL et al. Signal sequence within Fc{gamma}RIIA controls calcium wave propagation patterns: Apparent role in phagolysosome fusion. Proc Natl Acad Sci USA 2003.

125. Sawyer DW, Sullivan JA, Mandell GL. Intracellular free calcium localization in neutrophils during phagocytosis. Science 1985; 230(4726):663-666.

126. Lundqvist-Gustafsson H, Gustafsson M, Dahlgren C. Dynamic ca(2+)changes in neutrophil phagosomes A source for intracellular ca(2+)during phagolysosome formation? Cell Calcium 2000; 27(6):353-362.

127. Gagnon E, Duclos S, Rondeau C et al. Endoplasmic reticulum-mediated phagocytosis is a mechanism of entry into macrophages. Cell 2002; 110(1):119-131.

128. Harris HE, Weeds AG. Plasma gelsolin caps and severs actin filaments. FEBS Lett 1984; 177(2):184-188.

129. Serrander L, Skarman P, Rasmussen B et al. Selective inhibition of IgG-mediated phagocytosis in gelsolin-deficient murine neutrophils. J Immunol 2000; 165(5):2451-2457.

130. King WG, Mattaliano MD, Chan TO et al. Phosphatidylinositol 3-kinase is required for integrin-stimulated AKT and Raf-1/mitogen-activated protein kinase pathway activation. Mol Cell Biol 1997; 17(8):4406-4418.

131. Raeder EM, Mansfield PJ, Hinkovska-Galcheva V et al. Syk activation initiates downstream signaling events during human polymorphonuclear leukocyte phagocytosis. J Immunol 1999; 163(12):6785-6793.

132. Breton A, Descoteaux A. Protein kinase C-alpha participates in FcgammaR-mediated phagocytosis in macrophages. Biochem Biophys Res Commun 2000; 276(2):472-476.

133. Kamm KE, Stull JT. Dedicated myosin light chain kinases with diverse cellular functions. J Biol Chem 2001; 276(7):4527-4530.

134. Mansfield PJ, Shayman JA, Boxer LA. Regulation of polymorphonuclear leukocyte phagocytosis by myosin light chain kinase after activation of mitogen-activated protein kinase. Blood 2000; 95(7):2407-2412.

135. Olazabal IM, Caron E, May RC et al. Rho-kinase and myosin-II control phagocytic cup formation during CR, but not FcgammaR, phagocytosis. Curr Biol 2002; 12(16):1413-1418.

136. Chavrier P. May the force be with you: Myosin-X in phagocytosis. Nat Cell Biol 2002; 4(7):E169-171.

137. Diakonova M, Bokoch G, Swanson JA. Dynamics of cytoskeletal proteins during Fcgamma receptor-mediated phagocytosis in macrophages. Mol Biol Cell 2002; 13(2):402-411.

138. Ryder MI, Niederman R, Taggart EJ. The cytoskeleton of human polymorphonuclear leukocytes: Phagocytosis and degranulation. Anat Rec 1982; 203(3):317-327.

139. Stendahl OI, Hartwig JH, Brotschi EA et al. Distribution of actin-binding protein and myosin in macrophages during spreading and phagocytosis. J Cell Biol 1980; 84(2):215-224.

140. Valerius NH, Stendahl O, Hartwig JH et al. Distribution of actin-binding protein and myosin in polymorphonuclear leukocytes during locomotion and phagocytosis. Cell 1981; 24(1):195-202.

141. Allen LH, Aderem A. A role for MARCKS, the alpha isozyme of protein kinase C and myosin I in zymosan phagocytosis by macrophages. J Exp Med 1995; 182(3):829-840.

142. Swanson JA, Johnson MT, Beningo K et al. A contractile activity that closes phagosomes in macrophages. J Cell Sci 1999; 112(Pt 3):307-316.

143. Gold ES, Underhill DM, Morrissette NS et al. Dynamin 2 is required for phagocytosis in macrophages. J Exp Med 1999; 190(12):1849-1856.

144. Araki N, Johnson MT, Swanson JA. A role for phosphoinositide 3-kinase in the completion of macropinocytosis and phagocytosis by macrophages. J Cell Biol 1996; 135(5):1249-1260.

145. Merrifield CJ, Feldman ME, Wan L et al. Imaging actin and dynamin recruitment during invagination of single clathrin-coated pits. Nat Cell Biol 2002; 4(9):691-698.

146. Tse SM, Furuya W, Gold E et al. Differential role of actin, clathrin, and dynamin in Fc gamma receptor-mediated endocytosis and phagocytosis. J Biol Chem 2003; 278(5):3331-3338.

147. McNew JA, Parlati F, Fukuda R et al. Compartmental specificity of cellular membrane fusion encoded in SNARE proteins. Nature 2000; 407(6801):153-159.

148. Hackam DJ, Rotstein OD, Sjolin C et al. v-SNARE-dependent secretion is required for phagocytosis. Proc Natl Acad Sci USA 1998; 95(20):11691-11696.

149. Hackam DJ, Botelho RJ, Sjolin C et al. Indirect role for COPI in the completion of FCgamma receptor-mediated phagocytosis. J Biol Chem 2001; 276(21):18200-18208.

150. Coppolino MG, Kong C, Mohtashami M et al. Requirement for N-ethylmaleimide-sensitive factor activity at different stages of bacterial invasion and phagocytosis. J Biol Chem 2001; 276(7):4772-4780.

151. Vieira OV, Botelho RJ, Rameh L et al. Distinct roles of class I and class III phosphatidylinositol 3-kinases in phagosome formation and maturation. J Cell Biol 2001; 155(1):19-25.

152. Desjardins M, Huber LA, Parton RG et al. Biogenesis of phagolysosomes proceeds through a sequential series of interactions with the endocytic apparatus. J Cell Biol 1994; 124(5):677-688.

153. Desjardins M, Celis JE, van Meer G et al. Molecular characterization of phagosomes. J Biol Chem 1994; 269(51):32194-32200.

154. Vieira OV, Bucci C, Harrison RE et al. Modulation of Rab5 and Rab7 recruitment to phagosomes by phosphatidylinositol 3-kinase. Mol Cell Biol 2003; 23(7):2501-2514.

155. Alvarez-Dominguez C, Barbieri AM, Beron W et al. Phagocytosed live Listeria monocytogenes influences Rab5-regulated in vitro phagosome-endosome fusion. J Biol Chem 1996; 271(23):13834-13843.

156. Jahraus A, Tjelle TE, Berg T et al. In vitro fusion of phagosomes with different endocytic organelles from J774 macrophages. J Biol Chem 1998; 273(46):30379-30390.

157. Scott CC, Cuellar-Mata P, Matsuo T et al. Role of 3-phosphoinositides in the maturation of Salmonella-containing vacuoles within host cells. J Biol Chem 2002; 277(15):12770-12776.

158. Simonsen A, Lippe R, Christoforidis S et al. EEA1 links PI(3)K function to Rab5 regulation of endosome fusion. Nature 1998; 394(6692):494-498.

159. Callaghan J, Simonsen A, Gaullier JM et al. The endosome fusion regulator early-endosomal autoantigen 1 (EEA1) is a dimer. Biochem J 1999; 338(Pt 2):539-543.

160. Cantalupo G, Alifano P, Roberti V et al. Rab-interacting lysosomal protein (RILP): The Rab7 effector required for transport to lysosomes. EMBO J 2001; 20(4):683-693.

161. Jordens I, Fernandez-Borja M, Marsman M et al. The Rab7 effector protein RILP controls lysosomal transport by inducing the recruitment of dynein-dynactin motors. Curr Biol 2001; 11(21):1680-1685.

162. Garin J, Diez R, Kieffer S et al. The phagosome proteome: Insight into phagosome functions. J Cell Biol 2001; 152(1):165-180.

163. Dermine JF, Duclos S, Garin J et al. Flotillin-1-enriched lipid raft domains accumulate on maturing phagosomes. J Biol Chem 2001; 276(21):18507-18512.

164. Ehlenberger AG, Nussenzweig V. The role of membrane receptors for C3b and C3d in phagocytosis. J Exp Med 1977; 145(2):357-371.

165. Zhou M, Todd 3rd RF, van de Winkel JG et al. Cocapping of the leukoadhesin molecules complement receptor type 3 and lymphocyte function-associated antigen-1 with Fc gamma receptor III on human neutrophils. Possible role of lectin-like interactions. J Immunol 1993; 150(7):3030-3041.

166. Kindzelskii AL, Yang Z, Nabel GJ et al. Ebola virus secretory glycoprotein (sGP) diminishes Fc gamma RIIIB-to-CR3 proximity on neutrophils. J Immunol 2000; 164(2):953-958.

167. Galon J, Gauchat JF, Mazieres N et al. Soluble Fcgamma receptor type III (FcgammaRIII, CD16) triggers cell activation through interaction with complement receptors. J Immunol 1996; 157(3):1184-1192.

168. Poo H, Krauss JC, Mayo-Bond L et al. Interaction of Fc gamma receptor type IIIB with complement receptor type 3 in fibroblast transfectants: Evidence from lateral diffusion and resonance energy transfer studies. J Mol Biol 1995; 247(4):597-603.

169. Krauss JC, PooH, Xue W et al. Reconstitution of antibody-dependent phagocytosis in fibroblasts expressing Fc gamma receptor IIIB and the complement receptor type 3. J Immunol 1994; 153(4):1769-1777.

170. Worth RG, Mayo-Bond L, van de Winkel JG et al. CR3 (alphaM beta2; CD11b/CD18) restores IgG-dependent phagocytosis in transfectants expressing a phagocytosis-defective Fc gammaRIIA (CD32) tail minus mutant. J Immunol 1996; 157(12):5660-5665.

171. Sehgal G, Zhang K, Todd RF et al. Lectin-like inhibition of immune complex receptor-mediated stimulation of neutrophils. Effects on cytosolic calcium release and superoxide production. J Immunol 1993; 150(10):4571-4580.

172. Zhou MJ, Brown EJ. CR3 (Mac-1, alpha M beta 2, CD11b/CD18) and Fc gamma RIII cooperate in generation of a neutrophil respiratory burst: Requirement for Fc gamma RIII and tyrosine phosphorylation. J Cell Biol 1994; 125(6):1407-1416.

173. Majima T, Minegishi N, Nagatomi R et al. Unusual expression of IgG Fc receptors on peripheral granulocytes from patients with leukocyte adhesion deficiency (CD11/CD18 deficiency). J Immunol 1990; 145(6):1694-1699.

174. Gresham HD, Graham IL, Anderson DC et al. Leukocyte adhesion-deficient neutrophils fail to amplify phagocytic function in response to stimulation. Evidence for CD11b/CD18-dependent and -independent mechanisms of phagocytosis. J Clin Invest 1991; 88(2):588-597.

175. Isakov N. ITIMs and ITAMs. The Yin and Yang of antigen and Fc receptor-linked signaling machinery. Immunol Res 1997; 16(1):85-100.

176. Muta T, Kurosaki T, Misulovin Z et al. A 13-amino-acid motif in the cytoplasmic domain of Fc gamma RIIB modulates B-cell receptor signalling. Nature 1994; 369(6478):340.

177. Ono M, Bolland S, Tempst P et al. Role of the inositol phosphatase SHIP in negative regulation of the immune system by the receptor Fc(gamma)RIIB. Nature 1996; 383(6597):263-266.

178. Cox D, Dale BM, Kashiwada M et al. A regulatory role for Src homology 2 domain-containing inositol 5'-phosphatase (SHIP) in phagocytosis mediated by Fc gamma receptors and complement receptor 3 (alpha(M)beta(2); CD11b/CD18). J Exp Med 2001; 193(1):61-71.

179. Kant AM, De P, Peng X et al. SHP-1 regulates Fcgamma receptor-mediated phagocytosis and the activation of RAC. Blood 2002; 100(5):1852-1859.

180. Binstadt BA, Billadeau DD, Jevremovic D et al. SLP-76 is a direct substrate of SHP-1 recruited to killer cell inhibitory receptors. J Biol Chem 1998; 273(42):27518-27523.

181. Maresco DL, Osborne JM, Cooney D et al. The SH2-containing 5'-inositol phosphatase (SHIP) is tyrosine phosphorylated after Fc gamma receptor clustering in monocytes. J Immunol 1999; 162(11):6458-6465.

182. Nakamura K, Malykhin A, Coggeshall KM. The Src homology 2 domain-containing inositol 5-phosphatase negatively regulates Fcgamma receptor-mediated phagocytosis through immunoreceptor tyrosine-based activation motif-bearing phagocytic receptors. Blood 2002; 100(9):3374-3382.

Complement Receptors, Adhesion, and Phagocytosis

Eric Brown

Recognition of potential pathogens by host cells involved in their destruction is the initial step in generation of sterilizing immunity after epithelial barriers have been penetrated by disease-causing bacteria, viruses, and eukaryotes. One mechanism for recognition is that professional phagocytes express plasma membrane receptors for pathogen molecules, the pathogen-associated molecular patterns, or PAMPS. However, the recognition mechanism is potentially "leaky" because of the selection pressure on pathogens to evolve alterations in their surface structures that allow them to escape direct recognition. As a result, metazoan hosts have in turn evolved mechanisms for enhanced pathogen recognition, the most complex and sophisticated of which is the system of adaptive immunity. Yet adaptive immunity to new pathogens is slow, and successful host defense requires a rapid response that can at least hold pathogen growth in check until adaptive immunity develops. As a result, a number of pathogen recognition mechanisms have arisen that broaden the spectrum of pathogen containment. These recognition mechanisms involve soluble extracellular proteins that bind to the surfaces of possible pathogens and are in turn recognized by receptors on phagocytic cells. This indirect recognition of the pathogen in response to deposition of soluble host proteins is called opsonization.

One of the primary molecules involved in the bridging between pathogen and host phagocyte is the complement component C3. When complement is activated, C3 is deposited covalently onto pathogen surfaces in the form of C3b. This C3b, and its derivative iC3b, can be recognized by receptors on all professional phagocytes.

The Complement Cascade

Complement is a sequential series of proenzymes present in blood, which can be activated in response to infection in three distinct ways. The "classical pathway" is activated by antibody deposition (Fig. 1). Recognition of pathogen bound antibody by the C1q component of the first component of complement (C1) leads to stimulation of the enzymatic activity of C1, with subsequent cleavage and activation of the next two downstream components of the pathway, C4 and C2. Activated, cleaved C4 and C2, also bound to the pathogen surface, together make an enzyme that will cleave C3, leading to opsonization. Obviously, this classical pathway of complement activation is only relevant when pathogen-specific antibody is present. It nonetheless can be relevant in nonimmune hosts because of the presence in all of us of "natural antibody", presumably made in response to organisms colonizing our mucosal surfaces, which may crossreact with invading pathogens.

Molecular Mechanisms of Phagocytosis, edited by Carlos Rosales. ©2005 Eurekah.com and Springer Science+Business Media.

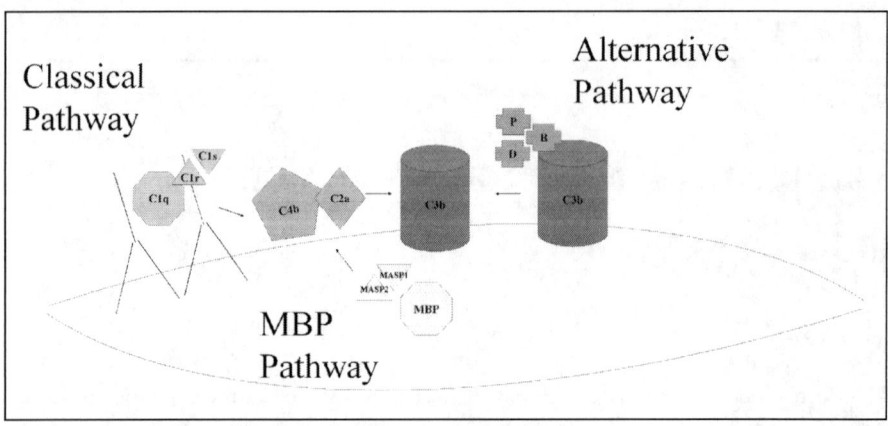

Figure 1. Three pathways of complement activation. The three pathways of complement activation are schematized. Antibody deposited on a bacterial surface can lead to activation of C1 of the classical pathway, and binding of mannose binding protein (MBP) to mannose terminated structures on the bacteria can activate MASP-1 and MASP-2. Both these pathways lead to activation and deposition of C4b2a, an enzyme that can cleave and activate C3, leading to deposition of C3b on the bacterial surface. C3b bound in this fashion or because of normal tickover of C3 can act as the nidus for assembly of Factors B, D, and P of the alternative pathway, which leads to an additional mechanism for activation and deposition of C3b on the bacterial surface. C3b and its derivative iC3b can be recognized by phagocyte complement receptors.

A second pathway for complement activation involves mannose binding protein (MBP), a plasma component that specifically recognizes high mannose carbohydrates, which often decorate the surfaces of invading pathogens. MBP is structurally related to C1q, the antibody recognition module of the classical pathway. When bound to pathogens, MBP can activate two enzymes, MASP-1 and MASP-2, which in turn cleave and activate C4 and C2. As in the classical pathway, this event leads to deposition of C3b and iC3b onto the pathogen surface.

The third pathway of complement activation by pathogens, termed the "alternative pathway", is quite distinct from the first two. This pathway takes advantage of the chemistry of the C3 component itself, which is continuously activated at a slow rate because of hydrolysis of a thioester bond (a phenomenon termed "tickover"), leading to nonspecific covalent deposition on nearby surfaces. Our own cells have multiple mechanisms for rapidly inactivating C3 deposited in this fashion so that it is not recognized either by phagocytic cells or other proteins of the complement activation cascade. In the absence of these control mechanisms, the deposited C3 can combine with other complement proteins, Factors B, D, and properdin, to rapidly cleave and deposit C3. Thus, on bacterial surfaces that lack mechanisms for control of the alternative pathway, C3 deposition is a highly potent amplification step for opsonization. Of course, C3 deposited by either the classical or the MBP pathway also can be amplified by the action of the alternative pathway, so that it has an important role in the accumulation of opsonic C3 fragments when these pathways are activated as well. The alternative pathway is especially important in filling the need for a broader recognition of potential invaders than possible through specific phagocyte PAMP receptors. For this pathway, active inhibition by host factors is required to prevent opsonization. There are pathogens that have discovered ways to prevent alternative pathway activation by mimicking host control proteins, but, in general, this pathway provides some protection against a very broad array of microbial invaders.

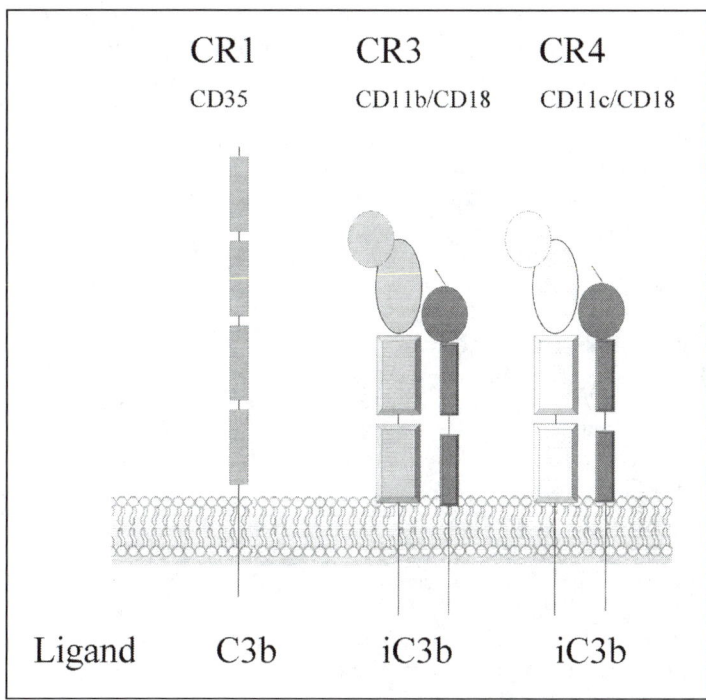

Figure 2. Complement receptors of phagocytic cells. The three major complement receptors of phagocytic cells are schematized. CR1, also known as CD35, is a type I membrane protein that binds C3b with higher affinity than iC3b. The integrin complement receptors CR3 and CR4 have the same β chain (CD18), but different α chains (CD11b and CD11c). These receptors recognize iC3b better than C3b.

Complement Receptors

Phagocytes express two distinct families of cell surface receptors for the deposited C3 fragments (Fig. 2). One receptor, CR1, is a member of the SCR family, characterized by multiple repeats of an approximately 60 aa motif, organized into 4 long homologous repeats. CR1 also may recognize the complement component C1q, which can bind to pathogens either directly or through the mediation of antibody opsonization,[1] and CR1 may recognize ligands outside the complement cascade that are structurally related to C1q.[2] CR1 is a type 1 membrane protein and is capable of phagocytosis under some circumstances. The CR1 cytoplasmic tail of ~60aa has no known motifs, so the mechanism by which it signals internalization remains uncertain, and little work on this topic has been reported.

Phagocytes also express receptors for deposited C3 that are members of the integrin family of cell adhesion receptors.[3] Integrins are heterodimeric cell surface molecules present on all cells that mediate binding of cells to each other and to the extracellular matrix. Ligand recognition within the family is quite diverse, and several integrins, including the C3 receptors, can recognize more than one ligand. The integrin heterodimers each contain one chain from a gene family known as α chains that are structurally related, and a second chain from a family of β chains. α chains and β chains are not related to each other by sequence, but they must pair with each other to progress through the endoplasmic reticulum into the secretory pathway for eventual expression on the cell surface. Only certain pairs of α and β chains form sufficiently stable heterodimers to allow cell surface expression (reviewed in ref. 4). The integrin

C3 receptors, (known as αMβ2 and αXβ2 in integrin nomenclature; other names in Figure 2 recognize the iC3b form of deposited C3. They are members of a subfamily of integrins all of which contain the β2 chain. β2 (also known as CD18) is expressed essentially exclusively on leukocytes. A characteristic of this family of integrins is that the α chains all contain an additional domain near their aminotermini known as the I (for "inserted") domain. The I domain is required for binding to virtually all ligands of the integrins in the β2 subfamily. It has a structure very similar to a portion of the β chain (known as the "I-like" domain), also intimately involved in ligand binding. As will be discussed below, a characteristic of this I domain fold is the capacity for a conformational change that is a key component of integrin function.

Regulation of Integrin Function

One characteristic of β2 integrins is that in resting cells, these receptors have exceedingly low affinity for ligand and virtually no ability to mediate cell adhesion to ligand coated surfaces, including iC3b-opsonized bacteria. When phagocytes have been activated by any of a variety of proinflammatory peptides or lipids that can accumulate at sites of infection and inflammation (a process known as inside-out signaling), integrins gain the capacity to mediate adhesion to ligand-coated surfaces. Recent work has unraveled the molecular mechanisms involved in regulation of acquisition of ligand recognition by integrin receptors. While the length and topic of this review precludes a detailed discussion of the data, the current model is summarized as follows (Fig. 3). In its resting state, the integrin is in a bent conformation so that the ligand binding domain actually is quite close to the plasma membrane.[5] This bent conformation involves interactions between the aminoterminus of the integrin

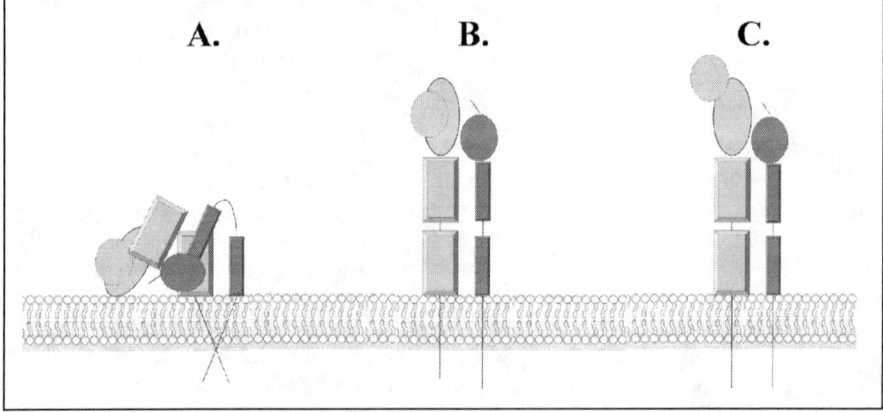

Figure 3. A model for affinity modulation of CD18 integrins. A) In the resting state, the integrins are in a bent conformation, with the interactions between the aminotermini and juxtamembrane sequences within the receptors that inhibit ligand binding. B) Activation as a result of inside-out signaling leads to separation of the cytoplasmic domains of the integrin chains, which propagates a conformational change through the membrane that releases the interactions between the aminotermini and juxtamembrane segments of the integrin. This leads to a straightening of the molecule that has been compared to the opening of a switchblade. C) The conformational changes in the integrin accompanying unbending leads to a further conformational change in the I domain at the aminoterminus of the molecule, which is the site of ligand binding. This conformationally altered I domain has increased ligand affinity because of greater accessibility of the ligand binding pocket to protein ligands.

and juxtamembrane sequences in the same molecule, and interaction with extracellular ligands is sterically inhibited. Upon activation, the integrin α and β chain cytoplasmic tails separate from each other,[6] leading to a conformational change through the plasma membrane that destabilizes the interaction of the integrin aminoterminus with the juxtamembrane domain, so that the integrin springs open, (the analogy has been made to the opening of a switchblade[7]), so that the aminoterminus is now facing out from the cell in a position much more favorable for extracellular ligand recognition. This major conformational change is transmitted to the α chain I domain (in the case of β2 integrins) and/or the I-like domain of the β chain, leading to an opening up of the actual ligand-binding cleft, which results in an increase in affinity for ligand. The extent to which the opening of the I domain affects ligand binding varies quite significantly among integrins. For αLβ2, the I domain conformational change increases ligand affinity by 3-4 orders of magnitude;[8] for the closely related αXβ2, the increase in affinity seems to be closer to 200 fold;[9] it may be even less for αMβ2.

In addition to the increased affinity of individual integrins, activation of integrin-mediated adhesion causes integrin clustering, resulting in an increase in net avidity. While the molecular mechanism linking conformational change to clustering is not well understood, recent data have led to the hypothesis that separation of the α and β chains, which initiates the process of conformational change in response to inside-out signaling, allows self association of β chain cytoplasmic domains,[10] which could lead in turn to integrin trimerization. Because of the complexity of currently available assay systems, it has been very difficult to sort out whether and in which circumstances alteration in the affinity of individual integrins or clustering is more important.

Inside-Out Signaling and Phagocytosis

As described above, integrins are adhesion receptors activated by allosteric regulation. What does this have to do with phagocytosis? Obviously, adhesion between phagocyte and phagocytic target is required for initiation of ingestion. The fact that integrins are designed to mediate intercellular adhesion makes them ideally suited for this purpose. The requirement for cell activation adds to their utility for recognition of complement opsonized targets, as well. As described above, because of C3 tickover in blood and extracellular fluid, a small amount of C3b and iC3b is generated continuously on all surfaces in contact with the blood. While there are rapid means for inactivating these deposited C3 fragments on our own cells, a ligand for phagocyte adhesion is nonetheless present transiently on our normal tissues. The requirement for cell activation prior to integrin receptor recognition of this ligand assures that nonspecifically deposited C3 will not be recognized by phagocytes. In the context of inflammation and infection, activators of inside-out signaling will be present that increase the affinity and avidity of the integrin recognition event, and alternative pathway activation will lead to increased deposition of complement opsonin, both leading to increased phagocyte recognition of an opsonized target. Synergy between these mechanisms for discrimination of background C3 deposition from opsonization of pathogens makes this a very sensitive step in recognition of potential agents of disease and helps limit leukocyte adhesion to sites of perturbation of homeostasis. In this context, it is important that ligation of IgG Fc receptors is a particularly potent mechanism for activation integrin-mediated adhesion. The presence of both antibody and complement on any surface is an excellent indication of infection. It has been know for many years that antibody and complement synergize to initiate phagocytosis of opsonized particles.[11] While this has often been interpreted as increased sensitivity to deposited IgG because of complement-mediated adhesion, the synergy also reflects activation of complement receptors by IgG ligation of Fc receptors.[12]

Complement Receptor Mediated Phagocytosis

Regulation of complement receptor mediated phagocytosis is more complex than simple activation of adhesion. This has been clear since the work of Silverstein almost 30 years ago, when he showed that peritoneal macrophages can bind complement opsonized targets but cannot phagocytose them without an additional signal.[13] Thus, activation of integrins to be adhesive is not sufficient to allow them to interact with the cellular machinery that mediates phagocytosis. Work by Griffin demonstrated that while complement receptors are not normally mobile in the plane of the membrane of unactivated cells, such mobility is required for phagocytosis.[14] Pharmacologic agents and cytokines that activate complement receptor mobility also activate complement-mediated phagocytosis. Based on modern understanding of integrin function, these data suggest that the resting macrophages used in these experiments likely expressed complement receptors that were of sufficient affinity to bind opsonized particles but were not able to cluster. Cell activation allowed the engaged receptors to cluster, a function required for phagocytosis. Among the characterized cytokines able to mediate this change in complement receptor function are IL-4, M-CSF, TNF-α, and GM-CSF.[15,16] Subsequently, several groups showed that activating signals could increase the diffusion of integrin receptors by releasing them from cytoskeletal constraint.[17-19] Based on these data, it is likely that affinity modulation of macrophage integrins is sufficient to allow binding of opsonized particles, but increased diffusion to allow clustering is required as well in order to activate ingestion. Thus, for complement-mediated phagocytosis, the molecular mechanisms that control integrin clustering represent a critical control point.

Much of the work on the molecular mechanisms involved in complement receptor mediated ingestion has centered on regulation of the cytoskeleton. The Arp2/3 complex, which nucleates branching actin polymerization to induce membrane protrusion, is required.[20] While activation of Arp2/3 dependent actin polymerization often is controlled by the Rho family GTPase Cdc42, RhoA itself seems to be more important in complement-mediated phagocytosis.[21,22] RhoA regulates phagocytosis because the Rho-activated kinase affects myosin function.[22] This contractility-dependent phagocytosis of complement-opsonized targets is consistent with morphologic observations that phagocytosis of these particles occurs through "sinking" into the cytoplasm rather than because of membrane protrusion and zippering, the classical mechanism for IgG-mediated ingestion.[23,24] Importantly, Rho family GTPases regulate integrin clustering rather than affinity.[25] Thus, it is likely that Rho activation leading to integrin clustering is upstream of a variety of alterations in the actin cytoskeleton that are required for complement-mediated ingestion.

Phorbol esters that activate protein kinase C (PKC) have been known for some years as stimuli that activate complement receptor-mediated phagocytosis.[26] As this would suggest, PKC is needed for complement mediated ingestion.[23] However, the isoforms of PKC involved and their targets in the ingestion process are not well understood. PKCε has been shown to be necessary for IgG-mediated phagocytosis,[27] but similar experiments have not been done for complement-dependent ingestion. PKCζ can regulate integrin clustering in lymphocytes[28] through a pathway that also involves RhoA; perhaps it also is involved in complement receptor clustering, which would explain its requirement in CR3-mediated ingestion.[29]

The nature of the initial cytoskeletal tethers that inhibit integrin diffusion and the mechanism of their loss in response to cell activation are less understood aspects of the regulation of complement-mediated ingestion. A role for the actin bundling protein L-plastin has been proposed, based on the ability of peptides from the aminoterminus of L-plastin to activate integrins, through both alteration of affinity and induction of integrin diffusion.[30,31] Integrin activation in myeloid cells is closely associated with L-plastin phosphorylation, suggesting the possibility that L-plastin phosphorylation regulates integrin activation through effects on actin crosslinking that can modulate integrin diffusion.

Cooperation between IgG Fc Receptors and Complement Receptors

As alluded to above, it has long been recognized that there is synergy between IgG and complement opsonization for phagocytosis.[11] One mechanism for synergy is that ligation of complement receptors enhances the efficiency of IgG binding to FcγRs on phagocytes; FcγRs are constitutively efficient at phagocytosis. At the same time, ligation of FcγRs can activate a signaling pathway that initiates a positive feedback loop, which leads to internalization of opsonized targets. In the absence of CR3, IgG-mediated ingestion is quite inefficient.[32] Enhancement of IgG-mediated ingestion by CR3 occurs even in the absence of complement opsonization. The nature of the ligand for CR3 in this circumstance and the molecular mechanism by which it enhances ingestion are both unclear. It is possible that CR3 can recognize carbohydrates on the ligated Fc receptor itself, leading to its recruitment to the site of phagocytosis.[33] This may facilitate association with the cytoskeleton,[34] a necessary early step in phagocytosis. There are additional mechanisms for cooperation between FcγRs and the integrin complement receptors as well. On human PMN, FcγRIIIB can physically associate with CR3.[35] Moreover, a complex between FcγRIIA, CR3, and the src family kinase hck has been found in PMN that have adhered via complement receptors.[36]

Integrin Regulation of Phagocytosis

A somewhat separate aspect of the function of integrins on phagocytes is that ligation of specific integrins can regulate the rate and extent of ingestion by Fc and C3 receptors.[37] Teleologically, this represents a mechanism for recognition by phagocytic cells that they are out of the bloodstream, at a site of infection or inflammation where their microbicidal capacity should be at its maximum. The enhanced phagocytosis activated by integrin ligation requires signaling between the ligated integrin and CR3, even when the opsonic ligand is IgG; thus, this mechanism for regulating phagocytosis has been called "integrin crosstalk". While several different integrins can activate crosstalk, αvβ3 appears to have a central role in this phenomenon. Even though this integrin is expressed minimally by PMN and many macrophages, antibody and peptide inhibition studies suggest that it has a major role in activation of CR3 by extracellular matrix ligands and even by phorbol esters.[38] The signaling pathway of integrin crosstalk is not completely understood. A second membrane molecule, known as Integrin Associated Protein (IAP; also, CD47) associates with αvβ3 and is required for initiation of signaling, since antibodies to IAP block this crosstalk, and crosstalk does not occur in IAP-deficient neutrophils.[39,40] IAP can activate heterotrimeric G protein signaling, and in PMN integrin crosstalk is pertussis toxin sensitive, suggesting that G protein signaling is involved in activating the phatocytic response. Neither the details of how heterotrimeric G proteins are activated in the apparent absence of a seven-transmembrane receptor nor the immediate downstream events in integrin crosstalk are known. In a model system, a serine in the β3 cytoplasmic tail is required for this signaling, which is manifest as inhibition of calmodulin-dependent kinase II.[41,42] Much remains to be understood of this pathway for regulation of phagocytosis.

Conclusion

Complement is a major opsonin in blood and in extracellular fluids that is a major mechanism for recognition and destruction of potential pathogen invaders. An important part of complement's role in this process is the recognition by phagocytes of complement components, particularly C3, bound to the surfaces of these dangerous organisms. Of the C3 receptors on the phagocytes, the two in the integrin family appear extremely important. In particular, the exquisite sensitivity of these integrin complement receptors to cues from the environment of the phagocytic cells help increase the efficiency with which they internalized opsonized targets for destruction and help limit their function to sites of inflammatory perturbation of homeostasis. Much has been learned about the molecular mechanisms involved in regulation of integrin receptor function, but many important events in the signal transduction cascades that lead to increased affinity for ligand and increased integrin clustering required for ingestion remain to be elucidated.

References

1. Eggleton P, Tenner AJ, Reid KBM. C1q receptors. Clin Exp Immunol 2000; 120(3):406-412.
2. Ghiran I, Barbashov SF, Klickstein LB et al. Complement receptor 1/CD35 is a receptor for mannan-binding lectin. J Exp Med 2000; 192(12):1797-1808.
3. Brown EJ. Complement receptors, adhesion, and phagocytosis. Infectious Agents and Disease J1 - IAD J2 - Inf Ag Dis 1992; 1:63-70.
4. Hynes RO. Integrins: Bidirectional, allosteric signaling machines. Cell 2002; 110(6):673-687.
5. Shimaoka M, Xiao T, Liu JH et al. Structures of the alpha L I domain and its complex with ICAM-1 reveal a shape-shifting pathway for integrin regulation. Cell 2003; 112(1):99-111.
6. Kim M, Carman CV, Springer TA. Bidirectional transmembrane signaling by cytoplasmic domain separation in integrins. Science 2003; 301(5640):1720-1725.
7. Takagi J, Petre BM, Walz T et al. Global conformational rearrangements in integrin extracellular domains in outside-in and inside-out signaling. Cell 2002; 110(5):599-511.
8. Shimaoka M, Lu C, Palframan RT et al. Reversibly locking a protein fold in an active conformation with a disulfide bond: Integrin alphaL I domains with high affinity and antagonist activity in vivo. Proc Natl Acad Sci USA 2001; 98(11):6009-6014.
9. Vorup-Jensen T, Ostermeier C, Shimaoka M et al. Structure and allosteric regulation of the alpha Xbeta 2 integrin I domain. PNAS 2003; 100(4):1873-1878.
10. Li R, Mitra N, Gratkowski H et al. Activation of integrin $\alpha IIb\beta 3$ by modulation of transmembrane helix associations. Science 2003; 300(5620):795-798.
11. Ehlenberger AG, Nussenzweig V. The role of membrane receptors for C3b and C3d in phagocytosis. J Exp Med 1977; 145(2):357-371.
12. Jones SL, Knaus UG, Bokoch GM et al. Two signaling mechanisms for activation of $\alpha M\beta 2$ avidity in polymorphonuclear neutrophils. Journal of Biological Chemistry 1998; 273:10556-10566.
13. Bianco C, Griffin Jr FM, Silverstein SC. Studies of the macrophage complement receptor. Alteration of receptor function upon macrophage activation. J Exp Med 1975; 141(6):1278-1290.
14. Griffin Jr FM, Mullinax PJ. Augmentation of macrophage complement receptor function in vitro. III. C3b receptors that promote phagocytosis migrate within the plane of the macrophage plasma membrane. J Exp Med 1981; 154(2):291-305.
15. Cross CE, Collins HL, Bancroft GJ. CR3-dependent phagocytosis by murine macrophages: Different cytokines regulate ingestion of a defined CR3 ligand and complement-opsonized Cryptococcus neoformans. Immunology 1997; 91(2):289-296.
16. Sampson LL, Heuser J, Brown EJ. Cytokine regulation of complement receptor-mediated ingestion by mouse peritoneal macrophages. M-CSF and IL-4 activate phagocytosis by a common mechanism requiring autostimulation by IFN-beta. J Immunol 1991; 146(3):1005-1013.
17. Kucik DF, Dustin ML, Miller JM et al. Adhesion activating phorbol ester increases the mobility of leukocyte integrin LFA-1 in cultured lymphocytes. 1996; 97:2139-2144.
18. Stewart MP, McDowall A, Hogg N. LFA-1-mediated adhesion is regulated by cytoskeletal restraint and by a Ca^{2+}-dependent protease, calpain. J Cell Biol 1998; 140(3):699-707.
19. Lub M, van Kooyk Y, van Vliet SJ et al. Dual role of the actin cytoskeleton in regulating cell adhesion mediated by the integrin lymphocyte function-associated molecule-1. Mol Biol Cell 1997; 8(2):341-351.
20. May RC, Caron E, Hall A et al. Involvement of the Arp2/3 complex in phagocytosis mediated by fcgammaR or CR3. Nat Cell Biol 2000; 2(4):246-248.
21. Caron E, Hall A. Identification of two distinct mechanisms of phagocytosis controlled by different Rho GTPases. Science 1998; 282(5394):1717-1721.

22. Olazabal IM, Caron E, May RC et al. Rho-kinase and myosin-II control phagocytic cup formation during CR, but not FcgammaR, phagocytosis. Curr Biol 2002; 12(16):1413-1418.

23. Allen LA, Aderem A. Molecular definition of distinct cytoskeletal structures involved in complement- and Fc receptor-mediated phagocytosis in macrophages. J Exp Med 1996; 184(2):627-637.

24. Kaplan G. Differences in the mode of phagocytosis with Fc and C3 receptors in macrophages. Scand J Immunol 1977; 6(8):797-807.

25. Schwartz MA, Shattil SJ. Signaling networks linking integrins and rho family GTPases. Trends Biochem Sci 2000; 25(8):388-391.

26. Wright SD, Silverstein SC. Tumor-promoting phorbol esters stimulate C3b and C3b' receptor-mediated phagocytosis in cultured human monocytes. J Exp Med 1982; 156(4):1149-1164.

27. Larsen EC, Ueyama T, Brannock PM et al. A role for PKC-epsilon in Fc gammaR-mediated phagocytosis by RAW 264.7 cells. J Cell Biol 2002; 159(6):939-944.

28. Giagulli C, Scarpini E, Ottoboni L et al. RhoA and zeta PKC control distinct modalities of LFA-1 activation by chemokines. Critical Role of LFA-1 Affinity Triggering in Lymphocyte In Vivo Homing. Immunity 2004; 20(1):25-35.

29. Lutz MA, Correll PH. Activation of CR3-mediated phagocytosis by MSP requires the RON receptor, tyrosine kinase activity, phosphatidylinositol 3-kinase, and protein kinase C zeta. J Leukoc Biol 2003; 73(6):802-814.

30. Jones SL, Wang J, Turck CW et al. A role for the actin-bundling protein L-plastin in the regulation of leukocyte integrin function. Proc Natl Acad Sci USA 1998; 95(16):9331-9336.

31. Wang J, Chen H, Brown EJ. L-plastin peptide activation of alpha(v)beta(3)-mediated adhesion integrin conformational change and actin filament disassembly. J Biol Chem 2001; 276(17):14474-14481.

32. Gresham HD, Graham IL, Anderson DC et al. Leukocyte adhesion deficient (LAD) neutrophils fail to amplify phagocytic function in response to stimulation: Evidence for CD11b/CD18-dependent and -independent mechanisms of phagocytosis. 1991; 88:588-597.

33. Zhou M, Todd IIIrd RF, van de Winkel JG et al. Cocapping of the leukoadhesin molecules complement receptor type 3 and lymphocyte function-associated antigen-1 with Fc gamma receptor III on human neutrophils. Possible role of lectin-like interactions. J Immunol 1993; 150(7):3030-3041.

34. Zhou MJ, Poo H, Todd RF et al. Surface-bound immune complexes trigger transmembrane proximity between complement receptor type 3 and the neutrophil's cortical microfilaments. J Immunol 1992; 148(11):3550-3553.

35. Petty HR, Worth RG, Todd IIIrd RF, Interactions of integrins with their partner proteins in leukocyte membranes. Immunol Res 2002; 25(1):75-95.

36. Zhou MJ, Brown EJ. CR3 (Mac-1α, Mβ2, CD11b/CD18) and Fc(gamma)RIII cooperate in generation of a neutrophil respiratory burst: Requirement for Fc(gamma)RII and tyrosine phosphorylation. Journal of Cell Biology 1994; 125:1407-1416.

37. Brown EJ, Lindberg FP. Matrix receptors of myeloid cells. In: Horton MA, ed. Blood Cell Biochemistry, volume 5: Macrophages and Related Cells. 1st ed. New York: Plenum Press, 1993:279-306.

38. Van Suijp JAG, Russell DG, Tuomanen E et al. Ligand specificity of 2purified complement receptor type 3 (CD11b/CD18, Mac-1, alphaM beta2): Indirect effects of an Arg-Gly-Asp sequence. 1993; 151:3324-3336.

39. Lindberg FP, Bullard DC, Caver TE et al. Decreased resistance to 22bacterial infection and granulocyte defects in IAP-deficient mice. Science 1996; 274:795-798.

40. Brown EJ, Frazier WA. Integrin-associated protein (CD47) and its ligands. Trends Cell Biol 2001; 11(3):130-135.

41. Blystone SD, Lindberg FP, LaFlamme SE et al. Integrin beta3 cytoplasmic tail is necessary and sufficient for regulation of α5β1 phagocytosis by αvβ3 and integrin-associated protein. Journal of Cell Biology 1995; 130:745-754.

42. Blystone SD, Slater SE, Williams MP et al. A molecular mechanism of integrin crosstalk: Alphavbeta3 suppression of Calcium/Calmodulin-dependent protein kinase II regulates alpha5beta1 function. Journal of Cell Biology 1999; 145(4):889-897.

Adding Complexity to Phagocytic Signaling:
Phagocytosis-Associated Cell Responses and Phagocytic Efficiency

Erick García-García and Carlos Rosales

Abstract

Regulation of the phagocytic process involves complex signaling pathways that lead to particle internalization and destruction. Phagocytosis, however, is not a cellular response occurring as an isolated event. Phagocytic signaling involves the regulation of many phagocytosis-associated cell responses that are important for host defense and for the resolution of the inflammatory process. In addition, due to the destructive nature of the phagocytic process, this cell response is tightly controlled, and phagocytes must respond to activation and differentiation signals that modulate their phagocytic efficiency. Presently it is unclear how all these events are coordinated. Available information, however, suggests a model in which phagocytosis-associated cell responses are regulated through signaling pathways that occur in parallel to, and partially overlap those regulating the ingestion process itself. Additionally, activation and differentiation signals appear to potentiate, or modify the utilization important signaling enzymes that regulate phagocytosis, in order to make this process more efficient.

Introduction

The immune system has a specialized subset of cells, named professional phagocytes, equipped for rapidly and efficiently ingesting invading microorganisms at sites of inflammation. These phagocytes are neutrophils and macrophages.[1] Monocytes (the macrophage precursors) are often included among the professional phagocytes, though they display a lower phagocytic response than neutrophils and macrophages.[1,2] Other nonprofessional phagocytes are dendritic cells,[3,4] fibroblasts,[5] microglia,[6] and some epithelial cell types.[7,8] An important difference between the phagocytic process of professional and nonprofessional phagocytes is that professional phagocytes respond to a phagocytic stimulus with many associated cell responses. These responses are of great importance in the context of the immune response, during tissue remodeling, and wound healing. Phagocytosis-associated cell responses include immuno-modulatory responses like the generation and release of pro-inflammatory[9,10] and anti-inflammatory mediators,[11] and also cell responses of destructive nature such as the respiratory burst,[12] and the release of toxic and microbicidal molecules by degranulation.[13,14] Additionally, in contrast to nonprofessional phagocytes, professional phagocytes are capable of recognizing a wide variety of phagocytic targets, and of ingesting them at a higher rate. However, in order to achieve controlled, yet efficient, phagocytic functions, professional phagocytes must

Molecular Mechanisms of Phagocytosis, edited by Carlos Rosales. ©2005 Eurekah.com and Springer Science+Business Media.

respond to activation and differentiation signals that modulate their phagocytic abilities. The regulation of phagocytosis-associated cell responses and the modulation of phagocytic efficiency thus represent a higher level of complexity for phagocytic signaling. Available evidence of how these events are regulated at the molecular level will be deal with in the next sections.

Regulation of Phagocytosis-Associated Cell Responses

As early as 1977 it was recognized that, in contrast to nonprofessional phagocytes, professional phagocytes were capable of secreting pro-inflammatory mediators in response to a phagocytic stimulus.[15] It is now well established that professional phagocytes respond to a phagocytic stimulus with a wide variety of associated cell responses.[1,2] Phagocytosis-associated cell responses appear to be either inflammatory or anti-inflammatory, predominantly depending on the properties of the phagocytic target. Inflammatory responses include the respiratory burst,[12] the release of pro-inflammatory mediators (cytokines, prostaglandins, and leukotrienes),[9,10] and the release of various microbicidal molecules by degranulation.[13,14] Anti-inflammatory responses are mainly characterized by the production and release of anti-inflammatory cytokines, and the active inhibition of the generation of pro-inflammatory mediators.[10]

The nature of the phagocytic target has a strong influence on the particular cell response that will accompany phagocytosis. Early reports indicated that phagocytosis of IgG-,[16-18] but not complement-opsonized[18,19] particles induced the release of granule enzymes and lipid-derived pro-inflammatory mediators. In contrast, phagocytosis of apoptotic cells by macrophages occurred in the absence of both responses.[20] More recently it was found that phagocytosis of apoptotic cells actually increases the secretion of the anti-inflammatory cytokine TGF-β.[21] Phagocytosis of necrotic cells by macrophages, on the other hand, induces an inflammatory phenotype characterized by the up-regulation of T cell costimulatory molecules, and enhanced ability for antigen presentation.[21] From these observations it follows that the receptors engaged in the phagocytic process will determine what associated cell responses will be triggered (Fig. 1). Phagocytosis mediated by immunoglobulin (Ig) receptors (FcRs) is associated with various inflammatory responses. Phagocytosis mediated by FcγRs (receptors for IgG) or FcαR (receptor for IgA) triggers pro-inflammatory cell responses including the respiratory burst, the release of microbicidal molecules by degranulation, and the production and release of lipid-derived pro-inflammatory factors and cytokines[2,22-25] (Fig. 1A). Phagocytosis mediated by complement receptors (CRs), on the other hand, is not always associated with inflammatory cell responses. Phagocytosis mediated by complement receptor 4 (CR4) occurs in association with the respiratory burst[26] (Fig. 1B). Complement receptor 3 (CR3), on the other hand, recognizes a wide variety of ligands,[27] and phagocytosis mediated by this receptor can occur in the absence or presence of the respiratory burst. This will depend on the ligand-binding site engaged on the receptor.[28] Engagement of the CR3 binding site for complement, during phagocytosis, does not trigger inflammatory responses.[18,19,26,29] In contrast, when the fibrinogen (a component of the coagulation cascade) binding site,[30] or the lectin site (that recognizes microorganism-derived sugars) on CR3[28,31] are engaged phagocytes present a strong respiratory burst and cytotoxic activity (Fig. 1C). In contrast to phagocytosis mediated by Fc receptors or complement receptors, phagocytosis of apoptotic cells has been directly associated with anti-inflammatory cell responses[11] (Fig. 1D). The anti-inflammatory nature of apoptotic cell phagocytosis may be fundamental for the clearance of apoptotic cells during tissue remodeling and wound healing. The engagement of the phosphatidylserine receptor during phagocytosis of apoptotic cells induces the production of anti-inflammatory mediators, and actively suppresses the production of inflammatory mediators.[5,32,33] MER is another receptor with a relevant role in the clearance of apoptotic cells,[34,35] and it was recently found that its stimulation is necessary for down-regulating inflammatory responses.[36,37] Interestingly, engagement of CR3 during phagocytosis of apoptotic cells also appears to down-regulate the production and secretion of

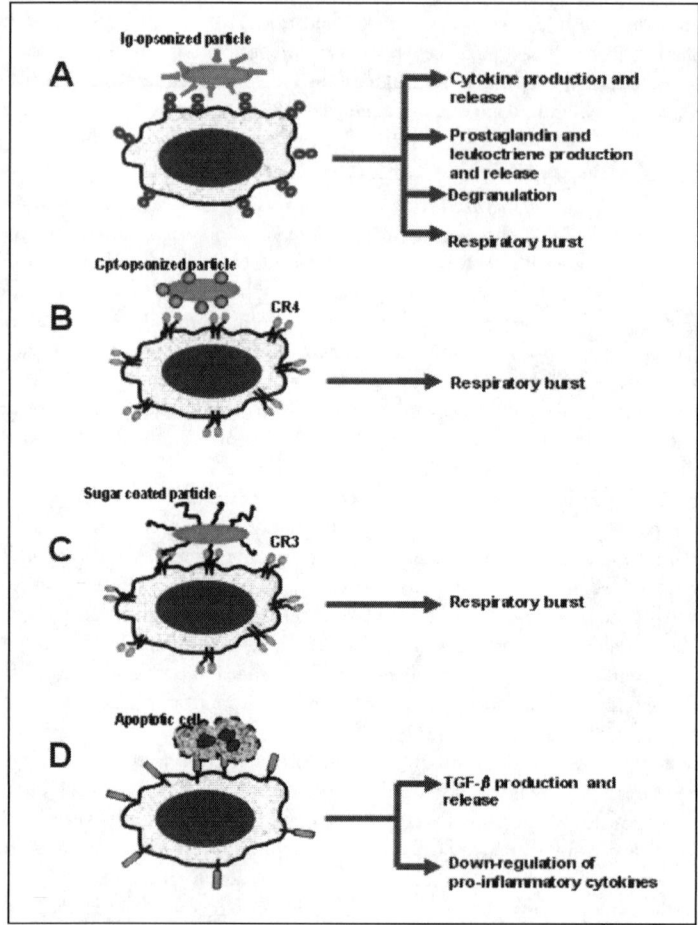

Figure 1. Triggering of phagocytosis-associated cell responses depends on the receptor engaged for phago-cytosis. Phagocytosis of immunoglobulin (Ig)-opsonized particles (A) triggers pro-inflammatory responses. Phagocytosis of complement (Cpt)-opsonized particles by CR4 (B), or by microorganism-derived sugars by CR3 (C) triggers the respiratory burst. Phagocytosis of apoptotic cells (D) induces the production of anti-inflammatory cytokines (such as TGF-β, transforming growth factor β), and down-regulates pro-inflammatory cytokine production.

pro-inflammatory cytokines.[38] Because CR3 may cooperate with the phosphatidylserine re-ceptor during phagocytosis of apoptotic cells,[39] it is possible that the anti-inflammatory effect of CR3 is due to simultaneous phosphatidylserine receptor stimulation.

It is currently unclear how these varied cell responses are regulated along with the phagocytic process. In some cases phagocytosis and accompanying cell responses appear to be regulated by parallel signaling pathways that are triggered by the same receptors that mediate phagocytosis (Fig. 2). For example, it has been shown that in monocytes FcγR-mediated phagocytosis is regulated through a signaling pathway dependent on PCK, and independent of PI 3-K and ERK[40] (Fig. 2). In the same cells, however, FcγR stimulation triggers a signaling pathway for cytokine production that depends on PI 3-K and ERK[40] (Fig. 2). In other cases phagocytosis and its associated cell responses appear to be regulated through partially overlapping signaling pathways (Fig. 3). In

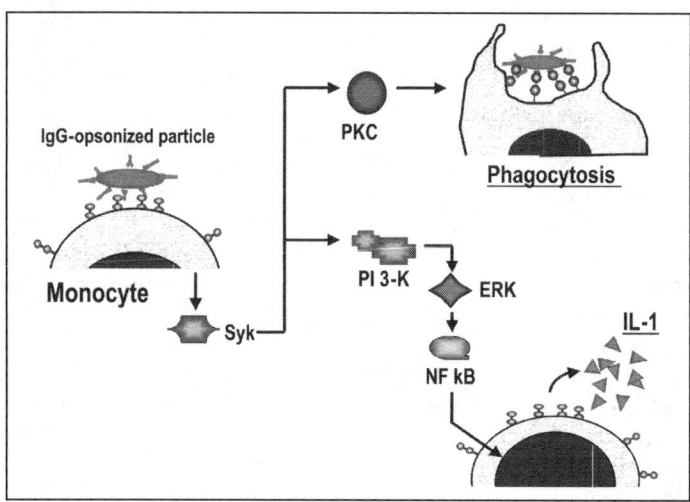

Figure 2. Phagocytosis and cytokine production are regulated by separate signaling pathways in monocytes. In monocytes stimulation of immunoglobulin G (IgG) receptors promotes phagocytosis and cytokine production. The signaling pathways that regulate these responses appear to be independent of one-another. PKC, protein kinase C; PI 3-K, phosphatidylinositol 3-kinase; ERK, extracellular signal-regulated kinase; NFκB, nuclear factor κB; IL-1, interleukine 1.

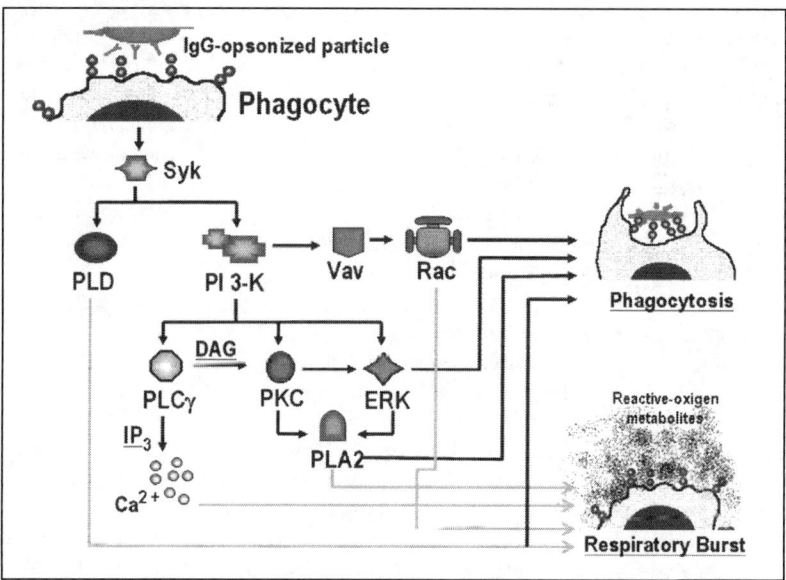

Figure 3. Phagocytosis and respiratory burst in professional phagocytes are regulated by partially overlapping signaling pathways. In neutrophils and macrophages stimulation of immunoglobulin G (IgG) receptors triggers phagocytosis and respiratory burst. The signaling pathways that regulate both responses share many signaling enzymes. However, phagocytosis, unlike the respiratory burst, is not regulated by intracellular calcium. See text for details. PKC, protein kinase C; PI 3-K, phosphatidylinositol 3-kinase; ERK, extracellular signal-regulated kinase; PLD, phospholipase D; PLCγ, phospholipase Cγ; PLA2, phospholipase A2; Ca^{2+}, intracellular calcium; DAG, diacylglycerol; IP3, inositol trisphosphate.

professional phagocytes, for instance, FcγR-mediated phagocytosis is regulated through a signaling pathway dependent on phosphatidylinositol 3-kinase (PI 3-K),[41,42] protein kinase C (PKC),[43,44] extracellular signal-regulated kinase (ERK),[40,45,46] Rac,[29] and various phospholipases[25] (Fig. 3). FcγR-mediated phagocytosis, however, can proceed independently of changes in the concentration of intracellular calcium[47,48] (Fig. 3). The activation of the respiratory burst, on the other hand, also depends PI 3-K,[49-51] PKC,[49,52] ERK,[53,54] Rac,[55,56] and various phospholipases,[50,57-59] but is in contrast dependent on calcium[60,61] (Fig. 3). In macrophages the activity of different PKC isoforms has been ascribed to the regulation of either phagocytosis, or the respiratory burst.[52] Phagocytosis appears to be regulated by PKCδ, and PKCε, while the respiratory burst is regulated by the PKCα isoform.[52] This observations suggest that, although the signaling pathways for phagocytosis and the respiratory burst share key signaling enzymes, they may not overlap completely.

Modulation of Phagocytic Efficiency

Compared to nonprofessional phagocytes, professional phagocytes present a great ability to recognize phagocytic targets, and ingest them at a higher rate.[1,2] In addition to this, professional phagocytes are able to modulate their phagocytic functions in response to activation and differentiation signals.

Cell Activation and Phagocytic Signaling

Phagocytosis efficiency of IgG- and complement-opsonized targets is up-regulated in response to a wide variety of activating stimuli (Fig. 4). These include, bacterial peptides,[62,63] cytokines,[64-68] lipid-derived pro-inflammatory mediators,[64,69,70] and the simultaneous stimulation of additional adhesive or phagocytic receptors.[71-74] How these activation signals up-regulate phagocytic efficiency at the molecular level is still unclear, however, in some cases the activating stimulus appears to potentiate the activation of signaling enzymes with important roles in phagocytosis (Fig. 4). For example, neutrophil activation with leukotriene B4 (a lipid-derived pro-inflammatory mediator) results in enhanced Syk activation.[70] This event correlates with increased phagocytic efficiency towards IgG-opsonized targets,[70] which is a cell response regulated by Syk.[75] Similarly, other activating stimuli have been reported to activate enzymes that professional phagocytes require for efficient phagocytosis, particularly ERK and PI 3-K.[42] Activating stimuli known to activate ERK and/or PI 3-K include, the bacterial peptide fMLP,[76,77] leukotrienes,[78,79] and cytokines such as IL-8,[80] granulocyte colony-stimulating factor,[81] and granulocyte-macrophage colony-stimulating factor.[77]

Cell Differentiation and Phagocytic Signaling

In the immune system neutrophils and macrophages represent the fully differentiated phagocytes. While neutrophils leaving the bone marrow are fully differentiated, macrophages differentiate from circulating monocytes in extra-vascular tissues.[82] Monocytes display a lower phagocytic response, compared to neutrophils and macrophages, and must respond to activation and differentiation signals in order to achieve optimal phagocytic capacity[1,2] (Fig. 5). It has been shown that the monocyte ability to ingest various targets is influenced by their state of differentiation. IgG-coated targets are ingested at a higher rate by mature macrophages, compared to undifferentiated monocytes[42] (Figs. 5, 6). Similarly, the ability to phagocytose via complement receptors is dependent on monocyte maturation.[83,84] While monocytes require an activating stimulus for internalizing complement-coated targets[83] (Fig. 5), mature macrophages ingest complement-opsonized targets in the absence of additional stimulus[83] (Fig. 6). Similarly, the ability of monocytes and bone marrow-derived macrophages to recognize and ingest apoptotic cells is inducible, and depends on their state of differentiation[85,86] (Figs. 5, 6). While the process of monocyte-to-macrophage differentiation has

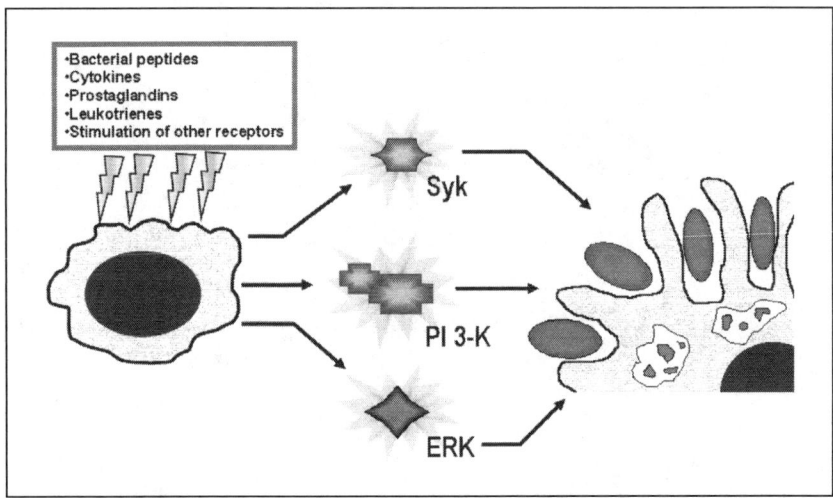

Figure 4. Phagocyte activation results in an enhanced phagocytic response. Phagocyte activation by various inflammatory stimuli potentiates the activities of signaling enzymes with important roles in regulating the phagocytic process. This activation enhances the phagocytic response towards various phagocytic targets. PI 3-K, phosphatidylinositol 3-kinase; ERK, extracellular signal-regulated kinase.

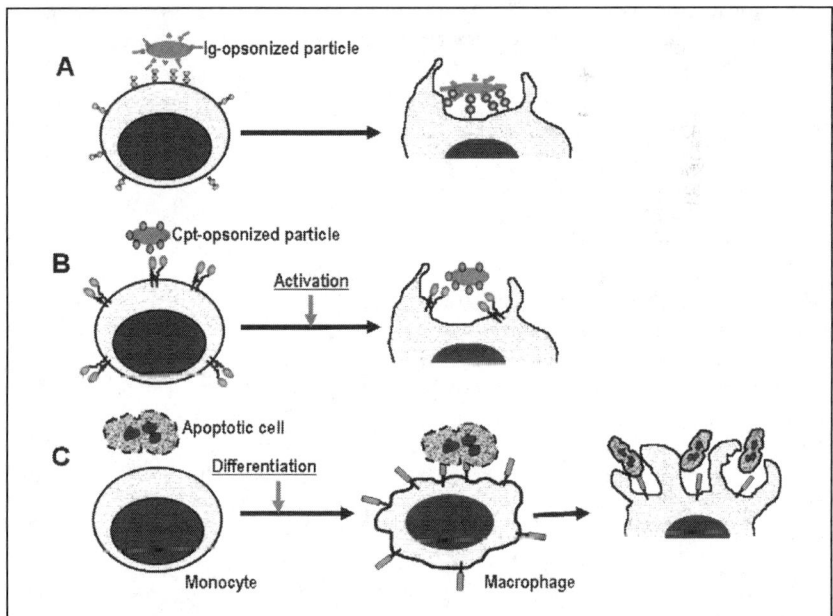

Figure 5. Recognition and ingestion of phagocytic targets by monocytes. Monocytes recognize and ingest immunoglobulin (Ig)-opsonized particles at low levels (A). Phagocytosis of complement (Cpt)-opsonized particles does not occur, unless the cell is activated by an additional stimulus (B). Monocytes fail to recognize and internalize apoptotic cells (C). This phagocytic function is developed upon monocyte-to-macrophage differentiation (C).

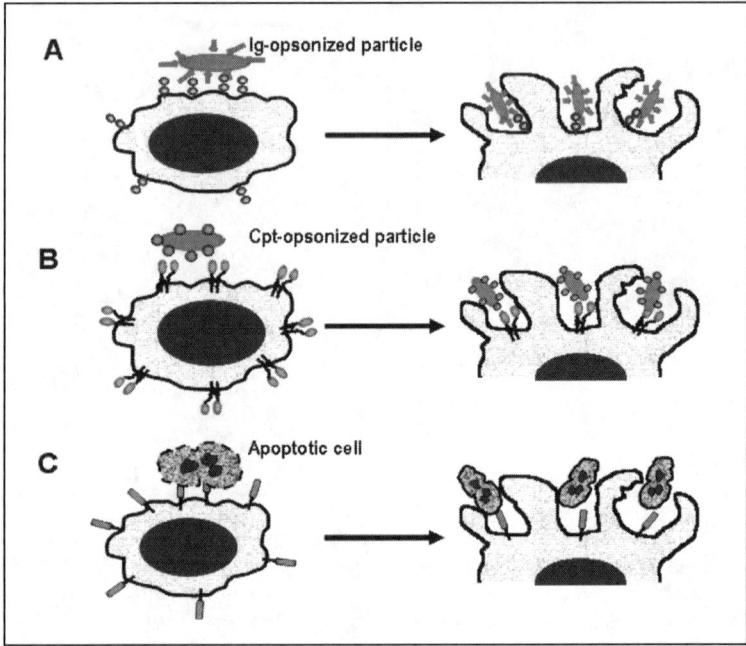

Figure 6. Recognition and ingestion of phagocytic targets by macrophages. Macrophages recognize and ingest very efficiently immunoglobulin (Ig)-opsonized targets (A), complement (Cpt)-opsonized targets (B), and also apoptotic cells (C).

been relatively well characterized,[9,82,87] the way the differentiation process enhances phagocytic ability is still unclear. Recent work, however, suggests that the differentiation process involves changes in the molecular machinery that regulates phagocytosis (Figs. 7, 8). In monocytes, the signaling pathway for phagocytosis depends on PKC,[42] but is independent of PI 3-K and ERK[40] (Fig. 7). The process of monocyte-to-macrophage differentiation, however, involves the ordered recruitment of PI 3-K and ERK for phagocytosis regulation.[42] Thus in mature macrophages phagocytosis of IgG-coated targets is regulated by PKC, and also by PI 3-K and ERK (Fig. 8). The PKC isoforms required for phagocytosis, however, appear to be different in monocytes and mature macrophages. In monocytes it was found that FcγRI stimulation increases PKC activity corresponding to the membrane translocation of the PKC isoforms δ, ε, and ζ[88] (Fig. 7). In monocyte-differentiated macrophages, on the other hand, FcγRI stimulation leads to PKC activity that corresponds to membrane translocation of the PKC isoforms α, β, and γ[88,89] (Fig. 8). It has also been demonstrated that PKC activation in monocyte-differentiated macrophages results in a stronger phagocytic response, compared to that of undifferentiated monocytes.[42] It is thus possible that by recruiting PI 3-K and ERK for phagocytosis regulation, and by regulating which PKC isoforms are recruited to the membrane, professional phagocytes achieve optimal levels of phagocytosis (Figs. 7, 8). Phospholipase A2 (PLA2) is another signaling molecule whose enzymatic activity is necessary for phagocytosis regulation. This enzyme acts down-stream of PKC and ERK, and the product of its enzymatic activity (arachidonic acid) regulates membrane remodeling events, that are necessary for pseudopod extension during phagocytosis.[90,91] During phagocytosis arachidonic acid is released through the enzymatic activity of various PLA2 isoforms,[25] and the release of this product also appears to be subjected to differentiation-dependent modulation. In monocytes

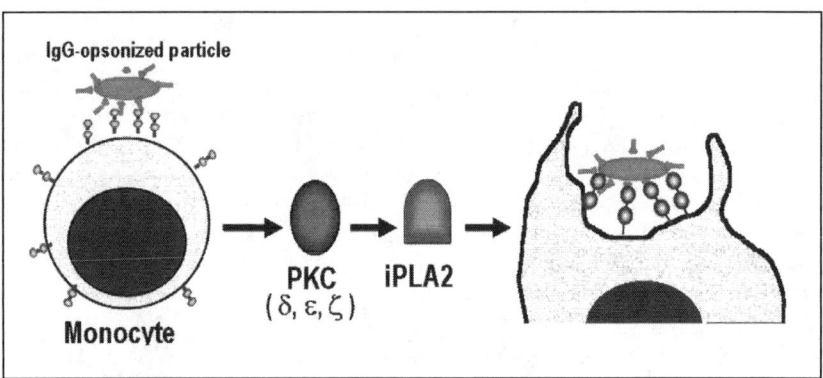

Figure 7. Signaling pathway for phagocytosis in monocytes. Monocytes ingest immunoglobulin G (IgG)-opsonized particles through a signaling pathway dependent on the protein kinase C (PKC) isoforms δ, ε, and ξ, and on calcium-independent phospholipase A2 (iPLA2).

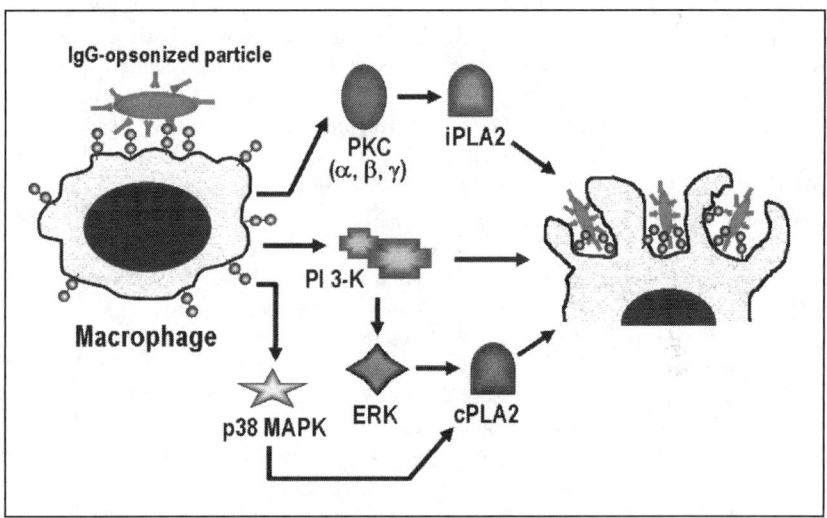

Figure 8. Signaling pathway for phagocytosis in professional phagocytes. Professional phagocytes ingest immunoglobulin G (IgG)-opsonized particles through a signaling pathway dependent on phosphatidylinositol 3-kinase (PI 3-K), extracellular signal-regulated kinase (ERK), p38 mitogen-activated protein kinase (p38MAPK), the PKC isoforms α, β and γ, and on calcium-independent (iPLA2), and calcium dependent (cPLA2) phospholipases A2.

arachidonic acid is released solely by calcium independent-PLA2, under PKC control[44,92,93] (Fig. 7). In macrophages, on the other hand, arachidonic acid is released by calcium dependent-PLA2, under ERK and p38 MAPK control,[94,95] and also by a calcium-independent PLA2, under PKC control[96] (Fig. 8). It is thus likely that by modulating the signaling pathways leading to arachidonic acid release, professional phagocytes also achieve an optimal phagocytic function. It thus appears that the process of monocyte-to-macrophage differentiation involves changes in the utilization of key signaling enzymes, whose activity may result in an enhanced phagocytic function.

Conclusion

Phagocytosis is cell response with important functions in the course of an immune response, but also during tissue remodeling and wound healing. Professional phagocytes are specialized cells capable of performing this function with great efficiency. The phagocytic process, however, is not a cell response that occurs as an isolated event, and a phagocytic stimulus triggers associated cell responses of destructive nature, but also immuno-modulatory cell responses. The concurrent regulation of these cell responses with phagocytosis sometimes involves separate, apparently independent signaling pathways. In other cases, however, although the signaling pathways for phagocytosis and associated responses share key signaling enzymes, these pathways may not be completely overlapping.

Professional phagocytes, in order to achieve controlled, yet efficient, phagocytic functions, must respond to activation and differentiation signals that modulate their phagocytic abilities. On one hand, activating stimulus appears to potentiate the activation of signaling enzymes with important roles in phagocytosis. On the other hand, differentiation signals that control the process of monocyte-to-macrophage differentiation promote a change in the utilization of key signaling enzymes for phagocytosis. This changes thus result in an enhanced phagocytic function.

The regulation of phagocytosis-associated cell responses and the modulation of phagocytic efficiency in response to activation and differentiation signals thus represent a higher level of complexity for phagocytic signaling. Although much work remains to be done in this field, future studies will enable us to better understand the complex signaling networks that regulate fundamental cellular mechanisms of the immune response.

Definitions

Apoptotic cell—Cell that has undergone the process of programmed cell death (apoptosis). This process is the result of the activation of specific signaling pathways, relevant during embryonic development, during the course of the immune response, and during the wound-healing process.

Complement—Family of soluble blood proteins that are part of the innate defense system. These proteins circulate in an inactive form, but become activated by the recognition of molecular components of microorganisms. Activation of complement components promotes their binding to the surface of microorganisms and function as opsonins. Additionally, activation of the complement cascade culminates in the formation of the membrane attack complex, that is capable of forming pores on the membrane of cells and pathogens.

Cytokines—Soluble, hormone-like proteins produced by leukocytes that act as a messengers between cells. Cytokines can stimulate or inhibit the growth and activity of various immune cells, and have effects on the differentiation and proliferation of the hematopoietic stem cells, as well as on the activation of lymphocytes and phagocytes.

Degranulation—Process whereby cells eject the contents of cytoplasmic vesicles (granules), either into a phagosome, or to the exterior of the cell through exocytosis. This process is characteristic of mast cells, basophils, neutrophils, macrophages, eosinophils, and platelets. During this process pro-inflammatory mediators, microbicidal and potentially cytotoxic substances are released from preformed cytoplasmic granules.

Nonprofessional phagocytes—Various cell types with low phagocytic activity. Nonprofessional phagocytes have more limited recognition and ingestion capabilities, compared to professional phagocytes.

Opsonization—Process whereby soluble proteins (such as immunoglobulins or complement components) bind to the surface pathogens and make them susceptible to ingestion by phagocytes.

Professional phagocytes—Specialized subset of cells in the immune system, capable of ingesting cells and pathogens with great efficiency, through the recognition of a wide array of molecular determinants of microorganisms, and also opsonins bound to pathogens and dying cells.

Prostaglandins and leukotrienes—A family of biologically active lipid compounds, derived from arachidonic acid by oxidative metabolism through the cyclooxygenase (prostaglandins), or 5-lipoxygenase (leukotrienes) pathways. They participate in host defense reactions modulating the inflammatory response. They have potent actions on many essential organs and systems, including the cardiovascular, pulmonary, and central nervous system, as well as the gastrointestinal tract and the immune system. Prostaglandins and leukotrienes are released by phagocytes.

Respiratory burst—Increase in oxidative metabolism of phagocytes following uptake of opsonized particles. The activation of the respiratory burst results in the production of toxic reactive-oxygen and nitrogen metabolites with bactericidal properties. These metabolites include hydrogen peroxide, hypohalites and nitric oxide.

Acknowledgements

Research is the author's laboratory is supported by grant 36407-M from Consejo Nacional de Ciencia y Tecnología (CONACyT), México.

References

1. Rabinovitch M. Professional and nonprofessional phagocytes: An introduction. Trends Cell Biol 1995; 5:85-87.
2. Jones SL, Lindberg FP, Brown EJ. Phagocytosis. In: Paul WE, ed. Fundamental Immunology. 4th ed. Philadelphia: Lippincott-Raven Publishers, 1999:997-1020.
3. Rubartelli A, Poggi A, Zocchi MR. The selective engulfment of apoptotic bodies by dendritic cells is mediated by the alpha(v)beta3 integrin and requires intracellular and extracellular calcium. Eur J Immunol 1997; 27:1893-900.
4. Albert ML, Pearce SF, Francisco LM et al. Immature dendritic cells phagocytose apoptotic cells via alphavbeta5 and CD36, and cross-present antigens to cytotoxic T lymphocytes. J Exp Med 1998; 188:1359-68.
5. Fadok VA, Bratton DL, Rose DM et al. A receptor for phosphatidylserine-specific clearance of apoptotic cells. Nature 2000; 405:85-90.
6. Witting A, Muller P, Herrmann A et al. Phagocytic clearance of apoptotic neurons by Microglia/ Brain macrophages in vitro: Involvement of lectin-, integrin-, and phosphatidylserine-mediated recognition. J Neurochem 2000; 75:1060-70.
7. Ryeom SW, Sparrow JR, Silverstein RL. CD36 participates in the phagocytosis of rod outer segments by retinal pigment epithelium. J Cell Sci 1996; 109(Pt 2):387-95.
8. Parnaik R, Raff MC, Scholes J. Differences between the clearance of apoptotic cells by professional and nonprofessional phagocytes. Curr Biol 2000; 10:857-60.
9. Gordon S, Clarke S, Greaves D et al. Molecular immunobiology of macrophages: Recent progress. Curr Op Immunol 1995; 7:24-33.
10. Rosenberg H, Gallin J. Inflammation. In: Paul WE, ed. Fundamental Immunology. 4th ed. Philadelphia: Lippincott-Raven Publishers, 1999:1051-1066.
11. Platt N, da Silva RP, Gordon S. Recognizing death: the phagocytosis of apoptotic cells. Trends Cell Biol 1998; 8:365-72.
12. Hampton M, Kettle A, Winterbourn C. Inside the neutrophil phagosome: Oxidants, myeloperoxidase and bacterial killing. Blood 1998; 92:3007-17.
13. Martin E, Ganz T, Lehrer RI. Defensins and other endogenous peptide antibiotics of vertebrates. J Leukoc Biol 1995; 58:128-36.
14. Lehrer RI, Ganz T. Antimicrobial polypeptides of human neutrophils. Blood 1990; 76:2169-81.
15. Bodel P, Miller H. Differences in pyrogen production by mononuclear phagocytes and by fibroblasts or HeLa cells. J Exp Med 1977; 145:607-17.

16. Nielsen OH, Elmgreen J, Thomsen BS et al. Release of leukotriene B4 and 5-hydroxyeicosatetraenoic acid during phagocytosis of artificial immune complexes by peripheral neutrophils in chronic inflammatory bowel disease. Clin Exp Immunol 1986; 65:465-71.
17. Brozna JP, Hauff NF, Phillips WA et al. Activation of the respiratory burst in macrophages. Phosphorylation specifically associated with Fc receptor-mediated stimulation. J Immunol 1988; 141:1642-7.
18. Stein M, Gordon S. Regulation of tumor necrosis factor (TNF) release by murine peritoneal macrophages: role of cell stimulation and specific phagocytic plasma membrane receptors. J Cell Sci 1991; 21:431-7.
19. Aderem AA, Wright SD, Silverstein SC et al. Ligated complement receptors do not activate the arachidonic acid cascade in resident peritoneal macrophages. J Exp Med 1985; 161:617-22.
20. Meagher L, Savill J, Baker A et al. Phagocytosis of apoptotic neutrophils does not induce macrophage release of thromboxane B2. J Leukoc Biol 1992; 52:169-73.
21. Barker R, Erwig L, Pearce W et al. Differential effects of necrotic or apoptotic cell uptake on antigen presentation by macrophages. Pathobiol 1999; 67:302-5.
22. Monteiro R, Van De Winkel J. IgA Fc receptors. Annu Rev Immunol 2003; 21:177-204.
23. Ravetch JV, Bolland S. IgG Fc receptors. Annu Rev Immunol 2001; 19:275-290.
24. Sánchez-Mejorada G, Rosales C. Signal transduction by immunoglobulin Fc receptors. J Leukoc Biol 1998; 63:521-533.
25. Lennartz MR. Phospholipases and phagocytosis: the role of phospholipid-derived second messengers in phagocytosis. Int J Biochem Cell Biol 1999; 31:415-430.
26. Berton G, Laudanna C, Sorio C et al. Generation of signals activating neutrophil functions by leukocyte integrins: LFA-1 and gp150/95, but not CR3, are able to stimulate the respiratory burst of human neutrophils. J Cell Biol 1992; 116:1007-17.
27. Ehlers MR. CR3: a general purpose adhesion-recognition receptor essential for innate immunity. Microbes Infect 2000; 2:289-94.
28. Le Cabec V, Cols C, Maridonneau-Parini I. Nonopsonic phagocytosis of zymosan and Mycobacterium kansasii by CR3 (CD11b/CD18) involves distinct molecular determinants and is or is not coupled with NADPH oxidase activation. Infect Immun 2000; 68:4736-45.
29. Caron E, Hall A. Identification of two distinct mechanisms of phagocytosis controlled by different Rho GTPases. Science 1998; 282:1717-21.
30. Rubel C, Fernandez GC, Rosa FA et al. Soluble fibrinogen modulates neutrophil functionality through the activation of an extracellular signal-regulated kinase-dependent pathway. J Immunol 2002; 168:3527-35.
31. Vetvicka V, Thornton BP, Ross GD. Soluble beta-glucan polysaccharide binding to the lectin site of neutrophil or natural killer cell complement receptor type 3 (CD11b/CD18) generates a primed state of the receptor capable of mediating cytotoxicity of iC3b-opsonized target cells. J Clin Invest 1996; 98:50-61.
32. Fadok VA, Bratton DL, Konowal ATrends in Cell Biology. Macrophages that have ingested apoptotic cells in vitro inhibit pro-inflammatory cytokine production through autocrine/paracrine mechanisms involving TGF-beta, PGE2, and PAF. J Clin Invest 1998; 101:890-8.
33. Huynh ML, Fadok VA, Henson PM. Phosphatidylserine-dependent ingestion of apoptotic cells promotes TGF-beta1 secretion and the resolution of inflammation. J Clin Invest 2002; 109:41-50.
34. Scott RS, McMahon EJ, Pop SM et al. Phagocytosis and clearance of apoptotic cells is mediated by MER. Nature 2001; 411:207-11.
35. Behrens EM, Gadue P, Gong SYTrends in Cell Biology. The mer receptor tyrosine kinase: expression and function suggest a role in innate immunity. Eur J Immunol 2003; 33:2160-7.
36. Camenisch T, Koller B, Earp HTrends in Cell Biology. A novel receptor tyrosine kinase, MER, inhibits TNF-alpha production and lipopolysacharide-induced endotoxic shock. J Immunol 1999; 162:3498-3503.
37. Cohen PL, Caricchio R, Abraham V et al. Delayed apoptotic cell clearance and lupus-like autoimmunity in mice lacking the c-mer membrane tyrosine kinase. J Exp Med 2002; 196:135-40.
38. Morelli AE, Larregina AT, Shufesky WJ et al. Internalization of circulating apoptotic cells by splenic marginal zone dendritic cells: dependence on complement receptors and effect on cytokine production. Blood 2003; 101:611-20.
39. Mevorach D, Mascarenhas JO, Gershov DTrends in Cell Biology. Complement-dependent clearance of apoptotic cells by human macrophages. J Exp Med 1998; 188:2313-20.

40. Garcia-Garcia E, Sanchez-Mejorada G, Rosales C. Phosphatidylinositol 3-kinase and ERK are required for NF-κB activation, but not for phagocytosis. J Leukoc Biol 2001; 70:649-658.
41. Cox D, Tseng CC, Bjekic G et al. A requirement for phosphatidylinositol 3-kinase in pseudopod extension. J Biol Chem 1999; 274:1240-1247.
42. Garcia-Garcia E, Rosales R, Rosales C. Phosphatidylinositol 3-kinase and extracellular signal-regulated kinase are recruited for Fc receptor-mediated phagocytosis during monocyte to macrophage differentiation. J Leukoc Biol 2002; 72:107-114.
43. Breton A, Descoteaux A. Protein kinase C-α participates in FcγR-mediated phagocytosis in macrophages. Biochem Biophys Res Com 2000; 276:472-476.
44. Karimi K, Gemmill TR, Lennartz MR. Protein kinase C and a calcium-independent phospholipase are required for IgG-mediated phagocytosis by Mono-Mac-6 cells. J Leukoc Biol 1999; 65:854-862.
45. Mansfield PJ, Shayman JA, Boxer LA. Regulation of polymorphonuclear leukocyte phagocytosis by myosin light chain kinase after activation of mitogen-activated protein kinase. Blood 2000; 95:2407-2412.
46. Raeder EM, Mansfield PJ, Hinkovska-Galcheva V et al. Syk activation initiates downstream signaling events during human polymorphonuclear leukocyte phagocytosis. J Immunol 1999; 163:6785-6793.
47. Myers JT, Swanson JA. Calcium spikes in activated macrophages during Fcgamma receptor-mediated phagocytosis. J Leukoc Biol 2002; 72:677-84.
48. Edberg JC, Lin CT, Lau D et al. The Ca^{2+} dependence of human Fc gamma receptor-initiated phagocytosis. J Biol Chem 1995; 270:22301-22307.
49. Yamamori T, Inanami O, Nagahata H et al. Roles of p38MAPK, PKC and PI3-K in the signaling pathways of NADPH oxidase activation and phagocytosis in bovine polymorphonuclear leukocytes. FEBS Lett 2000; 467:253-258.
50. Liu J, Liu Z, Chuai S et al. Phospholipase C and phosphatidylinositol 3-kinase signaling are involved in the exogenous arachidonic acid-stimulated respiratory burst in human neutrophils. J Leukoc Biol 2003; 74:428-37.
51. Hirsch E, Katanaev VL, Garlanda C et al. Central role for G protein-coupled phosphoinositide 3-kinase gamma in inflammation. Science 2000; 287:1049-53.
52. Larsen EC, DiGennaro JA, Saito N et al. Differential requirement for classic and novel PKC isoforms in respiratory burst and phagocytosis in RAW 264.7 cells. J Immunol 2000; 165:2809-2817.
53. Downey GP, Butler JR, Tapper H et al. Importance of MEK in neutrophil microbicidal responsiveness. J Immunol 1998; 160:434-443.
54. Dewas C, Fay M, Gougerot-Pocidalo MA et al. The mitogen-activated protein kinase extracellular signal-regulated kinase 1/2 pathway is involved in formyl-methionyl-leucyl-phenylalanine-induced p47phox phosphorylation in human neutrophils. J Immunol 2000; 165:5238-44.
55. Kreck ML, Freeman JL, Abo A et al. Membrane association of Rac is required for high activity of the respiratory burst oxidase. Biochemistry 1996; 35:15683-92.
56. Knaus UG, Heyworth PG, Evans T et al. Regulation of phagocyte oxygen radical production by the GTP-binding protein Rac 2. Science 1991; 254:1512-5.
57. Giron-Calle J, Srivatsa K, Forman HJ. Priming of alveolar macrophage respiratory burst by H(2)O(2) is prevented by phosphatidylcholine-specific phospholipase C inhibitor Tricyclodecan-9-yl-xanthate (D609). J Pharmacol Exp Ther 2002; 301:87-94.
58. Watson F, Lowe GM, Robinson JJ et al. Phospholipase D-dependent and -independent activation of the neutrophil NADPH oxidase. Biosci Rep 1994; 14:91-102.
59. Aebischer CP, Pasche I, Jorg A. Nanomolar arachidonic acid influences the respiratory burst in eosinophils and neutrophils induced by GTP-binding protein. A comparative study of the respiratory burst in bovine eosinophils and neutrophils. Eur J Biochem 1993; 218:669-77.
60. Bei L, Hu T, Qian ZM et al. Extracellular Ca2+ regulates the respiratory burst of human neutrophils. Biochim Biophys Acta 1998; 1404:475-83.
61. Hoyal CR, Gozal E, Zhou H et al. Modulation of the rat alveolar macrophage respiratory burst by hydroperoxides is calcium dependent. Arch Biochem Biophys 1996; 326:166-71.
62. Ogle JD, Noel JG, Sramkoski RM et al. Effects of chemotactic peptide f-Met-Leu-Phe (FMLP) on C3b receptor (CR1) expression and phagocytosis of microspheres by human neutrophils. Inflammation 1990; 14:337-53.

63. Rosales C, Brown EJ. Two mechanisms for IgG Fc-receptor-mediated phagocytosis by human neutrophils. J Immunol 1991; 146:3937-44.

64. Ogle JD, Noel JG, Sramkoski RM et al. The effects of cytokines, platelet activating factor, and arachidonate metabolites on C3b receptor (CR1, CD35) expression and phagocytosis by neutrophils. Cytokine 1990; 2:447-55.

65. Collins HL, Bancroft GJ. Cytokine enhancement of complement-dependent phagocytosis by macrophages: synergy of tumor necrosis factor-alpha and granulocyte-macrophage colony-stimulating factor for phagocytosis of Cryptococcus neoformans. Eur J Immunol 1992; 22:1447-54.

66. Pitrak D. Effects of granulocyte colony-stimulating factor and granulocyte-macrophage colony-stimulating factor on the bactericidal functions of neutrophils. Corr Opin Hematol 1997; 4:183-90.

67. Detmers PA, Powell DE, Walz A et al. Differential effects of neutrophil-activating peptide 1/IL-8 and its homologues on leukocyte adhesion and phagocytosis. J Immunol 1991; 147:4211-7.

68. Mitchell GB, Albright BN, Caswell JL. Effect of interleukin-8 and granulocyte colony-stimulating factor on priming and activation of bovine neutrophils. Infect Immun 2003; 71:1643-9.

69. Mancuso P, Peters-Golden M. Modulation of alveolar macrophage phagocytosis by leukotrienes is Fc receptor-mediated and protein kinase C-dependent. Am J Respir Cell Mol Biol 2000; 23:727-33.

70. Canetti C, Hu B, Curis J et al. Syk activation is a leukotriene B4-regulated event involved in macrophage phagocytosis of IgG-coated targets but not apoptotic cells. Blood 2003; 102:1877-83.

71. Arora M, Muñoz E, Tenner A. Identification of a site on mannan-binding lectin crititcal for enhancement of phagocytosis. J Biol Chem 2001; 276:43087-94.

72. Ohkuro M, Ogura-Masaki M, Kobayashi K et al. Effect of iC3b binding to immune complexes upon the phagocytic response of human neutrophils: synergistic functions between Fc gamma R and CR3. FEBS Lett 1995; 373:189-92.

73. Schnitzler N, Haase G, Podbielsky A et al. A costimulatory signal through ICAM-beta 2 integrin-binding potentiates neutrophil phagocytosis. Nat Med 1999; 5:231-5.

74. Rubel C, Fernandez GC, Dran G, Bompadre MB, Isturiz MA, Palermo MS. Fibrinogen promotes neutrophil activation and delays apoptosis. J Immunol 2001; 166:2002-10.

75. Kiefer F, Brumell J, Al-Alawi N et al. The Syk protein tyrosine kinase is essential for Fcgamma receptor signaling in macrophages and neutrophils. Mol Cell Biol 1998; 18:4209-20.

76. Chang LC, Wang JP. Examination of the signal transduction pathways leading to activation of extracellular signal-regulated kinase by formyl-methionyl-leucyl-phenylalanine in rat neutrophils. FEBS Lett 1999; 454:165-8.

77. Kodama T, Hazeki K, Hazeki O et al. Enhancement of chemotactic peptide-induced activation of phosphoinositide 3-kinase by granulocyte-macrophage colony-stimulating factor and its relation to the cytokine-mediated priming of neutrophil superoxide-anion production. Biochem J 1999; 337 (Pt 2):201-9.

78. Woo CH, You HJ, Cho SH et al. Leukotriene B(4) stimulates Rac-ERK cascade to generate reactive oxygen species that mediates chemotaxis. J Biol Chem 2002; 277:8572-8.

79. Ito N, Yokomizo T, Sasaki T et al. Requirement of phosphatidylinositol 3-kinase activation and calcium influx for leukotriene B4-induced enzyme release. J Biol Chem 2002; 277:44898-904.

80. Knall C, Worthen GS, Johnson GL. Interleukin 8-stimulated phosphatidylinositol-3-kinase activity regulates the migration of human neutrophils independent of extracellular signal-regulated kinase and p38 mitogen-activated protein kinases. Proc Natl Acad Sci USA 1997; 94:3052-7.

81. Kuroki M, O'Flaherty JT. Extracellular signal-regulated protein kinase (ERK)-dependent and ERK-independent pathways target STAT3 on serine-727 in human neutrophils stimulated by chemotactic factors and cytokines. Biochem J 1999; 341 (Pt 3):691-6.

82. Gordon S. Macrophages and the immune response. In: Paul WE, ed. Fundamental Immunology. Philadelphia: Lippincott-Raven Publishers, 1999:533-547.

83. Newman SL, Devery-Pocius JE, Ross GD et al. Phagocytosis by human monocyte-derived macrophages. Independent function of receptors for C3b (CR1) and iC3b (CR3). Complement 1984; 1:213-27.

84. Newman SL, Becker S, Halme J. Phagocytosis by receptors for C3b (CR1), iC3b (CR3), and IgG (Fc) on human peritoneal macrophages. J Leukoc Biol 1985; 38:267-78.

85. Fadok VA, Voelker DR, Campbell PA et al. The ability to recognize phosphatidylserine on apoptotic cells is an inducible function in murine bone marrow-derived macrophages. Chest 1993; 103:102S.

86. Newman SL, Henson JE, Henson PM. Phagocytosis of senescent neutrophils by human monocyte-derived macrophages and rabbit inflammatory macrophages. J Exp Med 1982; 156:430-42.

87. Valledor AF, Borras FE, Cullell-Young M et al. Transcription factors that regulate monocyte/macrophage differentiation. J Leukoc Biol 1998; 63:405-17.

88. Melendez AJ, Harnett MM, Allen JM. Differentiation-dependent switch in protein kinase C isoenzyme activation by FcγRI, the human high-affinity receptor for Immunoglobulin G. Immunol 1999; 96:457-464.

89. Melendez AJ, Harnett MM, Allen JM. FcγRI activation of phospholipase Cγ1 and protein kinase C in dibutyryl cAMP-differentiated U937 cells is dependent solely on the tyrosine-kinase activated form of phosphatidylinositol 3-kinase. Immunol 1999; 98:1-8.

90. Lennartz MR, Brown EJ. Arachidonic acid is essential for IgG Fc receptor-mediated phagocytosis by human monocytes. J Immunol 1991; 147:621-626.

91. Lennartz MR, Yuen AFC, McKenzie Masi S et al. Phospholipase A2 inhibition results in sequestration of plasma membrane into electronlucent vesicles during IgG-mediated phagocytosis. J Cell Sci 1997; 110:2041-2052.

92. Karimi K, Lennartz MR. Mitogen-activated protein kinase is activated during IgG-mediated phagocytosis, but it is not required for target ingestion. Inflammation 1998; 22:67-82.

93. Karimi K, Lennartz MR. Protein kinase C activation precedes arachidonic acid release during IgG-mediated phagocytosis. J Immunol 1995; 155:5786-5794.

94. Hazan-Halevy I, Seger R, Levy R. The requirement of both extracellular regulated kinase and p38 mitogen-activated protein kinase for stimulation of cytosolic phospholipase A2 activity by either FcγRIIA or FcγRIIIB in human neutrophils: A possible role of Pyk2 but not for the Grb-2-Sos-Shc complex. J Biol Chem 2000; 275:12416-12423.

95. Gijón MA, Spencer DM, Siddiqi AR et al. Cytosolic phospholipase A2 is required for macrophage arachidonic acid release by agonists that do and do not mobilize calcium: Novel role of mitogen-activated protein kinase pathways in cytosolic phosholipase A2 regulation. J Biol Chem 2000; 275:20146-20156.

96. Akiba S, Mizunaga S, Kume K et al. Involvement of group VI Ca^{2+}-independent phospholipase A2 in protein kinase C-dependent arachidonic acid liberation in zymosan-stimulated macrophage-like P388D$_1$ cells. J Biol Chem 1999; 274:19906-19912.

CHAPTER 6

Small GTP Binding Proteins and the Control of Phagocytic Uptake

Agnès Wiedemann, Jenson Lim and Emmanuelle Caron

Abstract

Phagocytosis is a conserved cellular process in Eukaryotes. A multi-step process, it involves the recognition of particulate material, e.g., microbes and apoptotic cells, their F-actin-driven engulfment and the subsequent destruction of the phagocytized material in phagolysosomes. Distinct sets of small GTP-binding proteins (Rap1, Arf6, Rho and Rab proteins) control and coordinate the successive steps of the phagocytic process. Moreover, these proteins are often targeted by microbial virulence factors. This review summarizes and discusses the evidence implicating Ras, Rho, Arf and Rab-family GTPases in the signalling pathways driving particle recognition and uptake.

Introduction

The last ten years have witnessed amazing progress in the understanding of the molecular basis of phagocytosis. Not only have the fundamental similarities between conventional macrophage phagocytosis and bacterial invasion of 'non-professional' phagocytes been recognized but high throughput technologies, new genetic systems[1] and the recent developments in cellular microbiology have contributed to draw the picture of a strong conservation in the mechanisms that underlie phagocytosis. The identification of new phagocytic receptors and signalling cascades has refined our understanding of the common features and individual variations amongst phagocytic pathways.

This review will focus on the role of small GTP-binding proteins in phagocytic uptake. Ras-like GTPases are a vast family of >100 molecular switches that control a plethora of essential aspects of cell biology, including cytoskeletal remodelling and vesicular trafficking, in a GTP-dependent manner. Their function, mediated by downstream effectors that bind specifically to the GTP-bound, active Ras proteins is tightly regulated by several classes of molecules: activators (Guanine nucleotide Exchange Factors, GEFs) and inactivators (GTPase-activating proteins, GAPs). Importantly, the relatively simple biochemistry of Ras-like GTPases has allowed the development of a variety of tools (e.g., epitope-tagged wild-type GTPases or mutant alleles, either dominant negative -thought to titrate endogenous GEFs- or constitutively active -thought to mimic the function of active GTP-binding proteins-) and assays (e.g., GTP-pull down assays, that use the GTPase-binding domain of specific downstream effectors to monitor the levels of active GTP-binding proteins in cells) for the analysis of GTPase function in cells.[2]

Molecular Mechanisms of Phagocytosis, edited by Carlos Rosales. ©2005 Eurekah.com and Springer Science+Business Media.

Rho Proteins and Actin Polymerisation

The most universal feature of phagocytic uptake is its actin-dependency and, indeed, actin polymerisation is the driving force that underlies both the zipper-like capture of phagocytic targets by phagocytes and the triggered intake of invasive bacteria by non professional phagocytes.[3] Unsurprisingly, like most other actin-dependent processes, phagocytosis generally requires the activity of Rho GTP-binding proteins, the evolutionarily conserved proteins that control cytoskeletal dynamics.[4] Therefore, whether a given phagocytic object is uptaken or not will depend primarily on how the phagocytic encounter affects Rho GTPase activity in host cells. However, which specific Rho family member(s) and regulators control actin polymerisation and uptake will vary.

Type I Phagocytosis (Rac/Cdc42)

The best understood phagocytic pathway links Fc receptors (FcR) on mammalian macrophages to the uptake of immunoglobulin (Ig)-opsonized targets. Referred to as type I phagocytosis, it involves the extension of pseudopods around the particle. FcR-mediated phagocytosis (mediated by either FcγR or FcεR), a known tyrosine kinase-dependent process, requires Cdc42 and Rac -but not Rho- activity, as proven using *Clostridium difficile* toxin B and overexpressed inhibitory constructs.[5-7] Rac and Cdc42 are also recruited to the site of particle binding during phagocytosis.[7] Moreover, phagocytic ligation of FcγR increases the levels of active Rac and Cdc42 in cells, suggesting that it activates these GTPases.[8-10] Finally, Rac and Cdc42 are activated independently of each other downstream of FcγR, respectively via the Rac GEF Vav and via an unknown Cdc42 exchange factor (Fig. 1).[8]

Likewise, Rac-1 and Cdc42 but not Rho control phagocytosis of *Neisseria gonorrhoae* by epithelial cells, after initial recognition of the adhesin/invasin Opa57 by CEACAM3, a member of the CEACAM (Carcinoma Embryonic Antigen-related Cellular Adhesion Molecules) family of receptors, which like phagocytic FcRs, contains an ITAM (Immunoreceptor Tyrosine Activated Motif) sequence in its cytosolic domain.[11] Similarly, interaction of Opa52-expressing *N. gonorrhoae* with human phagocytes through another CEACAM family member, CD66 requires Rac activity; in our opinion the possibility that Cdc42 is also involved has not been formally ruled out in this study.[12] Other bacteria bypass the need for receptor-induced activation of the Rho GTPases and inject bacterial effectors that will directly activate endogenous Rho GTPases in the host cell and thereby induce their uptake.[13] For example, *Salmonella typhimurium* induces a trigger-like uptake mechanism by injecting SopE/SopE2, two GEFs for Cdc42 and Rac in the target cell (Fig. 1).[14] Remarkably, these GEFs do not display the conventional Dbl homology domain that, in mammalian GEFs, is responsible for catalyzing nucleotide exchange.[15]

In mammalian cells, apoptotic cell (AC) uptake is another example of a Cdc42/Rac-dependent, Rho-independent process (Fig. 1).[16,17] GTP-pull down analysis has shown the reciprocal regulation of Rho and Rac activities in mammalian cells upon AC phagocytosis.[18] Unlike FcγR-mediated uptake, however, the activation of Rac induced upon AC binding is controlled by an unconventional bipartite exchange factor consisting of Dock180 and Elmo-1, that would be recruited to the site of ingestion by a Dock180-binding partner, CrkII.[19,20] A very similar signalling pathway operates in *C. elegans*, where the nematode orthologs of CrkII, Dock180, Rac and Elmo, called respectively CED-2, 5, 10 and 12 control apoptotic cell uptake.[21-24] Nevertheless, in the nematode as in other non-mammalian organisms, Rac -not Cdc42- seems to be the major regulator of phagocytic uptake. RacC controls phagocytosis in *Dictyostelium*,[25] while the use of SCAR mutants suggests a role for Rac in *Drosophila* phagocytosis.[26] SCAR is indeed a conserved regulator of actin polymerisation downstream of Rac.[27,28]

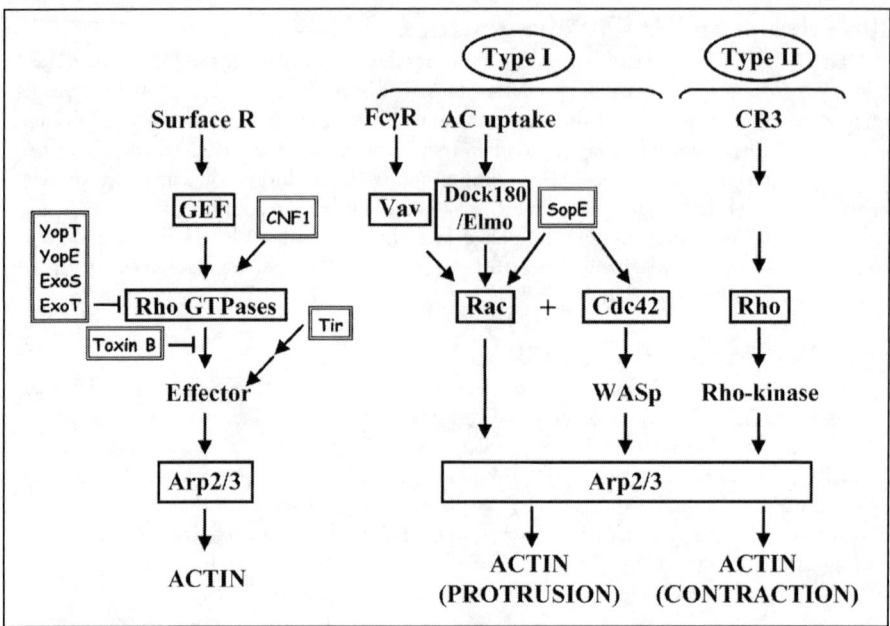

Figure 1. Activation of Rho, Rac and Cdc42 controls actin polymerization during phagocytosis. Left, a general model for phagocytic signalling. Phagocytosis generally involves the receptor-mediated, GEF-dependent activation of one or more Rho GTP-binding proteins, which will activate -through downstream effectors- the Arp2/3 complex and actin polymerisation. Several bacterial pathogens have the capacity to produce toxins or bacterial effectors (shown in double-lined boxes) that activate or inhibit the function or one or more Rho proteins, thereby modulating actin polymerization and phagocytosis. Right, the main signalling pathways activated during phagocytosis in mammalian cells. Type I phagocytosis is characterized by the independent activation of Rac and Cdc42, whereas only Rho activity is required during type II phagocytosis. GEF, guanine nucleotide exchange factor; CNF1, cytotoxic necrotizing factor 1 Tir, translocated intimin receptor; Yop, *Yersinia* outer protein; FcγR, Fcγ receptor; AC, apoptotic cell; CR3, complement receptor type 3; WASp, Wiskott-Aldrich Syndrome protein. See text for details.

Altogether, analysis of the phagocytic pathways that are associated with the formation of protrusions (type I) has revealed a common requirement for Cdc42 and Rac function in mammalian cells but only for Rac in *Dictyostelium*, an organism that lacks a clear Cdc42 homolog[29] and in *C. elegans*, where a Cdc42 homolog has been identified.[30] One may wonder what is the exact function of Cdc42 during mammalian phagocytosis as local activation of Rac, but not Cdc42 at the plasma membrane is sufficient to promote uptake.[31] Nonetheless, the main classes of protrusion-associated phagocytic pathways are regulated by Rac (plus Cdc42 in mammalian cells), with no apparent role for Rho in uptake, whilst they differ in the specific signalling pathways and GEFs that control Rac (and maybe Cdc42) activation (Fig. 1).

Type II Phagocytosis (Rho)

A second class of phagocytic events has been identified, that do not involve such a dramatic extension of protrusions.[32] Type II phagocytosis is mediated in mammals by the complement receptor 3 (CR3, Mac-1, CD11b/CD18, $\alpha_M\beta_2$), which signals to the actin cytoskeleton in a Cdc42/Rac-independent but Rho-dependent manner. Rho (but not Rac or Cdc42) is also recruited to forming CR3 phagosomes.[7] The molecular mechanisms underlying Rho recruitment and function at phagosomes are still poorly understood, although Rho kinase, a Rho effector is clearly mediating Rho effects on the cytoskeleton during phagocytosis (Fig. 1).[33]

G-protein-coupled receptor proteinase activated receptor-2 (PAR-2) mediates melanosome uptake in human keratinocytes in a Rho- and Rho kinase-dependent manner.[34] Whether phagocytosis through this receptor is the second known example of type II phagocytosis will have to await further investigation. In particular, the roles of Rac and Cdc42 should be tested in this model.

Variations and Anomalies

A number of other, opsonin-independent phagocytic encounters illustrate variations on the type I/Rac-Cdc42 and type II/Rho paradigm. Receptor-mediated uptake of the parasite *Leishmania amazonensis* amastigotes by Chinese hamster ovary cells is an actin-, tyrosine kinase-, PI3-kinase- and myosin-dependent process that is accompanied by activation of Cdc42 but not Rac and that requires Rho and Cdc42 but not Rac function.[35] Likewise, the activation of both Rho and Cdc42 but not Rac was shown to be necessary for the internalization of virulent encapsulated *N. meningitidis* following their type IV pilus-mediated adhesion to vascular endothelial cells.[36] In contrast to the trigger-like mechanism induced by *Salmonella typhimurium*, *Listeria monocytogenes* actively promotes its entry into mammalian cells by inducing a zipper-like uptake process.[37] *Listeria* entry into epithelial cells is therefore likely to resemble conventional phagocytosis and to involve a receptor-mediated, Rho GTPase-dependent uptake process, accompanied by GTPase recruitment and activation. In line with this, YopE -a *Yersinia* type III secretion system (TTSS) effector with GAP activity against Rho, Rac and Cdc42 (see below)- blocks *Listeria* uptake.[38] Moreover, in non phagocytic cells, the uptake of beads coated with internalin B -one of the two main surface expressed bacterial proteins that mediate *Listeria* entry- is independent of Rho but requires Rac (and in some cell lines also Cdc42) activity.[39] By contrast, all three prototypical Rho GTPases (Cdc42, Rac and Rho) control the trigger-like invasion of epithelial cells by *Shigella flexneri*.[40] Cdc42 and Rac are required for actin polymerization at the site of *Shigella* entry and expression of the TTSS effector IpaC in host cells is sufficient to induce Cdc42/Rac-mediated cytoskeletal changes.[41] More recently, it was demonstrated that VirA, a TTSS effector protein essential for *Shigella* invasion[42] binds tubulin, promotes microtubule destabilization, triggers Rac activation (through an unknown mechanism) and Rac1-dependent membrane ruffling.[43] Rho allows the transformation of the initial surface extensions into a more stable adhesive-like structure that is permissive for entry.[44] Rho, Rac and Cdc42 activities also control the uptake of *Brucella abortus* by human epithelial cells and invasin-mediated uptake of *Yersinia enterocolitica* by professional phagocytes.[45,46] However, adhesin-mediated, β_1 integrin-dependent invasion of epithelial cells by *Y. pseudotuberculosis*, a process accompanied by the recruitment of Rho, Rac and Cdc42 to the site of particle binding, only requires Rac function.[47,48] The origin of these discrepancies in *Yersinia*-induced signalling and uptake remains unclear.

Importantly, the central role of Rho GTPases in phagocytosis explains why they are targeted by so many bacterial toxins and virulence factors.[49,50] Toxins secreted by members of the *Clostridium* genus, and some strains of *Escherichia coli*, *Yersinia* and *Bordetella* modulate bacterial uptake by activating or inactivating specific sets of small GTPases, either from the Rho subfamily, or from both the Rho and Ras subfamilies.[50] Unsurprisingly, toxins that inhibit Rho activity block phagocytic uptake while the activating toxins promote phagocytosis (Fig. 1).[13,50] Likewise, several TTSS effectors directly perturb the Rho GTPase activation cycle. The *Yersinia* cystein protease, YopT, inactivates Rho, Rac and Cdc42 in vivo, by cleaving both the GTP- and GDP-bound forms of the protein and extracting them from membranes.[51-54] *Yersinia* YopE, and *Pseudomonas aeruginosa* ExoS and ExoT, display RhoGAP activity in vitro and can inactivate RhoA, Rac and Cdc42, but not Ras or Ral.[55-57] Therefore, these bacteria are able to inhibit their phagocytosis by forcing the conversion of Rho family members to their GDP-bound, inactive states, preventing them from activating downstream effectors and actin polymerisation.

In striking contrast to the examples listed above, microbial pathogens like enteropathogenic *E. coli* (EPECs), exert their early effects on the cytoskeleton not by acting on Rho GTPase cycling but rather by bypassing Rho proteins. When EPECs interact with epithelial cells, Tir is inserted into the host cell plasma membrane and directs the formation of actin-rich pedestals, in a Nck-, N-WASP- and Arp2/3-dependent manner.[58] However neither Rho, Rac, nor Cdc42 are required for pedestal formation. It is worth mentioning that one or more Rho GTPases are subsequently involved in EPEC invasion of epithelial cells, as it is a toxin B-sensitive process.[59] Finally, several protozoan parasites enter host cells independently of the host cell actin-dependent internalization machinery and Rho GTPase activity. Rather, they mobilize their own secretory organelles, such as the micronemes, rhoptries and dense granules of *Plasmodium* and *Toxoplasma*, in order to deliver microbial products at the host-pathogen interface and invade the host cell cytosol. For example, during *Trypanosoma cruzi* invasion, binding of an unidentified parasite product to host cells causes an increase in intracellular calcium, which destabilizes the cortical actin cytoskeleton and induces the microtubule-mediated recruitment and fusion of lysosomes to the plasma membrane.[60] No data is available on host cell Rho GTPases in this context, although they are obviously very unlikely to be involved. A very surprising result comes from another protozoan, *Entamoeba histolytica*, in which increased Rac activity has been correlated with inhibition of phagocytosis.[61]

Regardless of their Rho GTPase-dependency, most of the actin-driven phagocytic pathways we discussed converge onto the Arp2/3 complex to promote a local and oriented F-actin polymerization at the site of particle binding (Fig. 1). This is true for example for type I and type II phagocytosis[62,33] in mammals, and in *Dictyostelium*.[63] This is not surprising, as several mechanisms can account for Arp2/3 recruitment and activation, all relying on signal-induced exposure of the Arp2/3-binding, acidic domain of an adapter protein.[64] The best understood example of such an adapter is the Wiskott-Aldrich syndrome protein (WASp), a Cdc42 downstream effector that mediates activation of the Arp2/3 complex in a phosphatidylinositol 4,5-bisphosphate-dependent manner[65] and regulates type I phagocytosis in mammals.[66,67]

Ras Proteins and Receptor Activation

Whereas the evidence linking Rho GTPases to phagocytic uptake is compelling, the role of other small GTP-binding proteins is far less understood and often restricted to one receptor-mediated pathway. Several members of the Ras subfamily of GTPases have been linked with phagocytic uptake in mammalian and *Dictyostelium* cells. Importantly, several clostridial toxins can inactivate Ras proteins -in addition to Rac- and cause inhibition of phagocytosis as well as complete rounding of the cell body and detachment from the substratum.[68-70]

CR3, an integrin expressed at the surface of mammalian phagocytes and involved in phagocyte adhesion and motility mediates binding and type II-phagocytosis of particles opsonized with the complement fragment C3bi (see above).[71] CR3 is constitutively inactive in resting phagocytes and an activation signal is required to enable binding of most of the CR3 ligands.[72,73] Binding of C3bi-opsonised particles and subsequent CR3-mediated phagocytosis by activated macrophages are blocked by a dominant negative mutant of Rap1. Conversely, a constitutively active Rap1 mutant induces a two- to threefold increase in the number of C3bi-opsonised particles that bind to resting macrophages, thereby bypassing the need for an activation signal. Altogether, these results show the essential role of Rap1 in CR3 activation, a prerequisite step for the uptake of complement-opsonised particles.[72] However, the mechanisms controlling CR3 activation downstream of Rap1 are still unknown; in particular whether Rap1 controls a change in integrin affinity or avidity is unclear.[74] By contrast, Rap1 activity does not influence the binding of IgG-opsonised targets to FcγR[72], a process known to be constitutively active.[75] Although the requirement for Rap1 activation in mammalian phagocytosis is at the moment restricted to the $\alpha_M\beta_2$/CR3 receptor, the general role of this small GTP-binding protein in adhesion-related processes[76] makes it likely that other integrin-dependent phagocytic events mediated by mammalian phagocytes, such as $\alpha_V\beta_3$ and $\alpha_V\beta_5$-mediated apoptotic cell uptake[77-79] will be controlled by Rap1.

Interestingly, DdRap1, the *Dictyostelium discoideum* Rap1 homolog[80] is also essential for phagocytosis.[81] *D. discoideum* strains expressing constitutively active or wild type Rap1 internalise twice as many latex beads or *E. coli* bacteria as compared to the wild type strain. However, a strain expressing dominant negative Rap1 shows a 50% decrease in phagocytosis compared to the wild type strain. Although *D. discoideum* does not express integrins,[82] other adhesive receptors have been suggested to control phagocytosis, an EDTA-sensitive process in this organism. DdCad-1 (Ca^{2+}-dependent cell-cell adhesion molecule-1) and two nine-transmembrane receptors, Phg1 and SadA (Substrate adhesion-deficient A) all localise to phagosomes during uptake. However, only Phg1- and SadA- (not Dd-Cad-1)-deficient mutants show a phagocytic defect.[82-84] Therefore, the function of Rap1 could be a common, conserved requirement during phagocytosis through adhesion receptors, both in *Dictyostelium* and in mammalian cells.

Although first described in experiments designed to identify genes that could revert the phenotype of K-Ras- transformed fibroblasts,[85] Rap1 has since been found to control a complex set of functions, most of which relate to adhesion. In this respect, the regulation of CR3 receptor activation during macrophage phagocytosis is just one of the adhesion-related functions of Rap1.[76] Interestingly, in yeast cells, the Rap1 homolog, Bud1p accumulates at sites of polarized growth and budding and is involved in recruiting and activating Cdc24p, a Cdc42 GEF.[86,87] Rap1 could serve a similar, polarity-related function during phagocytosis, i.e., determine the direction of engulfment of particles, like a biological compass, and control the recruitment and activation of Rho-family GTPases : RacC in *Dictyostelium*[25,81] and Rho during CR3-mediated phagocytosis in mammalian macrophages.[7]

Unlike Rap1, the evidence linking Ras to phagocytosis is scarce and contradictory. On the one hand, when dominant negative H-Ras was microinjected into mouse macrophages, CR3-mediated phagocytosis was still observed,[72] suggesting that H-Ras activation plays no role in this process. On the other hand, the *Dictyostelium* H-Ras homolog, RasS was shown to regulate phagocytosis and cell motility: RasS-null cells show impaired fluid phase endocytosis and phagocytosis, but enhanced locomotion.[88] As mentionned above there is a strong correlation between adhesion and phagocytosis in *Dictyostelium*, where cell motility suppresses phagocytic cup formation.[88] Overall these results would suggest that there is no direct function for Ras in phagocytic uptake, although it does not exclude a role for Ras in other aspects of the phagocytic process (e.g., phagocytosis-induced transcriptional activation, phagocyte survival).

The other two Ras-family members that have been examined in the context of phagocytosis are R-Ras and RalA, although none of them appears to play a fundamental role. Expression of constitutively active R-Ras but not RalA leads to phagocytosis of C3bi-opsonised particles by resting macrophages.[89] However expression of dominant negative R-Ras or RalA in activated macrophages has no effect on CR3-mediated phagocytosis.[72] This suggests that R-Ras can promote CR3 activation, as reported for other integrins,[90-92] but is not required during CR3-dependent uptake.

Together, these results suggest a conserved albeit not universal role for a few Ras-subfamily proteins in phagocytosis. Rap1– and possibly Ras-control signalling pathways that enable adhesion-like receptors to perform their phagocytic function.

Arfs and Rabs and the Delivery of Membrane to Forming Phagosomes

It has been realised for a long time that phagocytosis would ultimately result in the net loss of plasma membrane if a mechanism ensuring membrane replenishment did not exist.[93,94] There is now strong evidence to suggest that particle uptake requires the delivery of endomembrane to forming phagosomes.[95,96] In all eukaryotic cells, vesicular trafficking between subcellular compartments and to the plasma membrane is controlled by Rab and Arf-subfamily GTPases.[2] So far, three of these small GTP-binding proteins have been implicated in phagocytic uptake. Interestingly, many more Rab proteins (e.g., Rab2-5, 7, 11, 14) were detected on mammalian phagosomes.[97] However some of them like Rab2 (involved in ER to Golgi and intraGolgi transport), Rab5 (early endosome fusion) and Rab7 (a late endosome marker) are more likely to be involved in the maturation -rather than the formation of phagosomes.[98,99]

In *Dictyostelium*, the activity of RabD, a close homolog of mammalian Rab14, conditions the rate of phagocytic uptake.[100] Whether these results are solely explained by a positive role for RabD, a marker of the endolysosomal and contractile vacuole systems, on membrane delivery to forming phagosomes, or whether overexpressed RabD mutants have also a general effect on the size or binding ability of *Dictyostelium* cells, remains unclear. Rab11 was also involved in phagocytosis, both in *Dictyostelium* and in mammalian phagocytes. However, inhibition of Rab11 activity has opposite effects on uptake in the two cell systems. In single cell amoebae, where Rab11 associates primarily with the contractile vacuole system and not with endosomes, expression of a dominant negative Rab11 allele doubles the rate of particle association to cells.[101] By contrast, in mammalian cells Rab11 regulates endosome recycling and Rab11 activity is required for optimal FcγR-mediated phagocytosis, as dominant negative Rab11 halves the number of IgG-opsonized red blood cells (RBCs) that are taken up.[102] These results are compatible with a model where Rab11 activity influences uptake by controlling the availability of distinct sources of intracellular membrane : the vacuolar system, which enlarges when dominant negative Rab11 is expressed in *Dictyostelium*, and the Rab11-controlled recycling endosome pool in macrophages. Finally Arf6, a known regulator of membrane trafficking, cell polarity, adhesion and motility[103] is the sole Arf-family member to be involved in phagocytic uptake. Indeed, phagocytosis is insensitive to brefeldin A, a fungal metabolite that inhibits most high molecular weight ArfGEFs and therefore the functions mediated by Arf1-5.[104] Whilst Arf6 gets activated following FcγR ligation and also recruited to early phagosomes, overexpression of dominant negative Arf6 mutants blocks FcγR-mediated uptake.[10,104] Although Arf6 was first thought to play a dual role in actin polymerization and membrane delivery during uptake, a recent study has shown that Arf6 function is essential for pseudopod extension around IgG-opsonized RBCs but dispensable for actin polymerisation (Fig. 2). Specifically, Arf6 controls the exocytosis of VAMP3-positive recycling endosomes at nascent FcγR phagosomes.[10]

A handful of studies have recently indicated that Arf6 and Rab GTPases play a role in uptake, by controlling the recruitment of membrane to extending pseudopodia. It will be interesting to see whether mammalian Rab14 and other Rab proteins (e.g., Rab3, a known regulator of exocytosis) also influence the delivery of membrane during phagocytosis. Another important question to address is the universality[105] and conservation of Arf6/Rab involvement in the different phagocytic systems that have become available. A more daunting task will be to understand both the signals that activate Arf6, Rab11 and possibly other Rabs and the mechanisms by which these GTP-binding proteins coordinate membrane recruitment during uptake.

Conclusions and Perspectives

The parallel analysis of the molecular basis of phagocytosis and bacterial pathogenesis has revealed a major role for several small GTP-binding proteins in regulating particle uptake. Remarkably, distinct GTPase subfamilies control successive steps of the phagocytic process (Fig. 2). First, a Ras-like GTPase, Rap1, controls the ability of activation-dependent adhesion

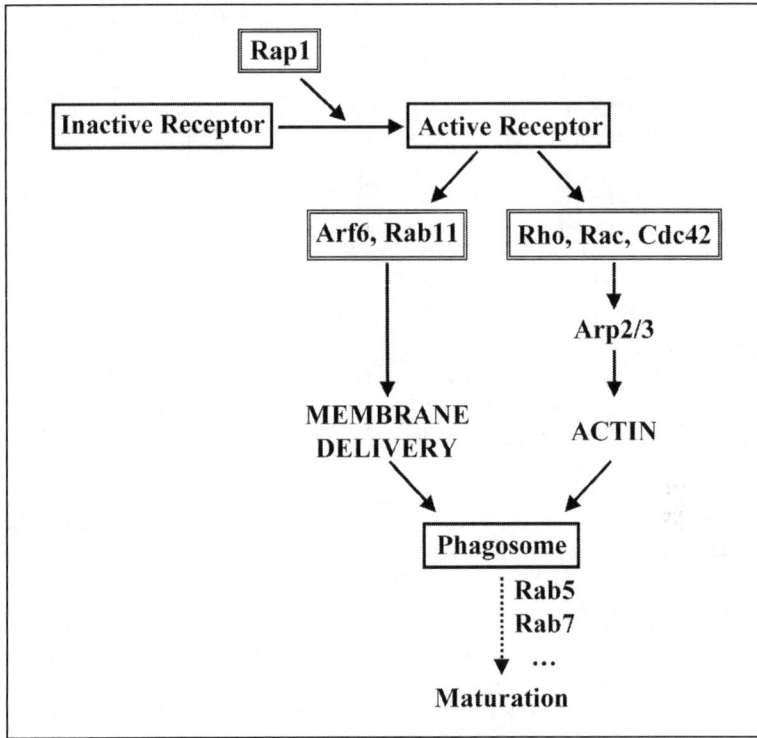

Figure 2. Distinct small GTPase subfamilies control different steps during phagocytosis. In this theoretical model, particle binding induces two GTPase-mediated cascades that coordinate phagocytic uptake by controlling actin polymerization and membrane delivery at the nascent phagosome. Rap1 controls the binding function of the adhesion-related phagocytic receptors that are activation-dependent. See text for details.

receptors (e.g., integrin-family members like CR3) to bind their ligand. Second, once particles have bound to host cells, Rho proteins control the remodelling of the actin cytoskeleton that drives uptake. Generally, particle binding triggers the local recruitment and activation of one or several Rho proteins, which mediate actin polymerisation in an Arp2/3-dependent manner. Which particular signalling pathway and Rho protein are activated during uptake will depend, however, on the nature of the initial receptor/ligand interaction at the cell surface. In line with their conserved, essential role in phagocytosis, Rho proteins and the pathways they control are frequently targeted by bacterial virulence factors and toxins, which thereby influence -positively or negatively- bacterial uptake. Third, optimal phagocytosis requires the delivery of endosome-derived membranes to forming phagosomes, a process controlled by an Arf-family member, Arf6 and by Rab proteins. The available evidence suggests that the GTPase-controlled pathways leading to membrane delivery and actin polymerisation are independently activated upon particle binding. How the distinct signalling cascades that mediate the different steps of the uptake process are activated and coordinated remains unknown. It is clear, however, that small GTP-binding proteins play a crucial role not only in phagocytic uptake but also in several responses associated to phagocytosis, such as the activation of the NADPH oxidase[106,107] and phagosome maturation (Fig.2). There is little doubt that the detailed analysis of small GTPase function during phagocytosis will continue to further our understanding of this essential cell function.

Definitions

Phagocytic uptake—A restricted view of phagocytosis, limited to the binding and internalisation of particulate material.

Small GTP-binding protein—Signalling molecule and molecular switch (a.k.a. small GTPase, Ras-subfamily member) that is active when bound to GTP and inactive when bound to GDP.

Downstream effector—Protein that binds specifically to the active, GTP-bound form of the small GTP-binding proteins and mediates its function.

Virulence factor—A product that contribute to the ability of a microbe to cause disease.

Toxin—A virulence factor that is secreted extracellularly.

Acknowledgements

The authors thank Céline Cougoule for critical reading of the manuscript. Research in the Caron laboratory is supported by grants from the Wellcome Trust and the BBSRC (Biotechnology and Biological Sciences Research Council).

References

1. Cardelli J. Phagocytosis and macropinocytosis in Dictyostelium: phosphoinositide-based processes, biochemically distinct. Traffic 2001; 2:311-320.
2. Takai Y, Sasaki T, Matozaki T. Small GTP-binding proteins. Physiol Rev 2001; 81:153-208.
3. Swanson JA, Baer SC. Phagocytosis by zippers and triggers. Trends Cell Biol 1995; 5:89-93.
4. Etienne-Manneville S, Hall A. Rho GTPases in cell biology. Nature 2002; 420:629-635.
5. Cox D, Chang P, Zhang Q et al. Requirements for both Rac1 and Cdc42 in membrane ruffling and phagocytosis in leukocytes. J Exp Med 1997; 186:1487-1494.
6. Massol P, Montcourrier P, Guillemot JC et al. Fc receptor-mediated phagocytosis requires Cdc42 and Rac1. EMBO J 1998; 17:6219-6229.
7. Caron E, Hall A. Identification of two distinct mechanisms of phagocytosis controlled by different Rho GTPases. Science 1998; 282:1717-1721.
8. Patel JC, Hall A, Caron E. Vav regulates activation of Rac but not Cdc42 during FcgammaR-mediated phagocytosis. Mol Biol Cell 2002; 13:1215-1226.
9. Kant AM, De P, Peng X, et al. SHP-1 regulates Fcgamma receptor-mediated phagocytosis and the activation of Rac. Blood 2002; 100:1852-1859.
10. Niedergang F, Colucci-Guyon E, Dubois T et al. ADP ribosylation factor 6 is activated and controls membrane delivery during phagocytosis in macrophages. J Cell Biol 2003; 161:1143-1150.
11. Billker O, Popp A, Brinkmann V, et al. Distinct mechanisms of internalization of Neisseria gonorrhoeae by members of the CEACAM receptor family involving Rac1- and Cdc42-dependent and -independent pathways. EMBO J 2002; 21:560-571.
12. Hauck CR, Meyer TF, Lang F et al. CD66-mediated phagocytosis of Opa52 Neisseria gonorrhoeae requires a Src-like tyrosine kinase- and Rac1-dependent signalling pathway. EMBO J 1998; 17:443-454.
13. Gruenheid S, Finlay BB. Microbial pathogenesis and cytoskeletal function. Nature 2003; 422:775-781.
14. Zhou D, Galan J. Salmonella entry into host cells: the work in concert of type III secreted effector proteins. Microbes Infect 2001; 3:1293-1298.
15. Buchwald G, Friebel A, Galan JE et al. Structural basis for the reversible activation of a Rho protein by the bacterial toxin SopE. EMBO J 2002; 21:3286-3295.
16. Leverrier Y, Ridley AJ. Requirement for Rho GTPases and PI 3-kinases during apoptotic cell phagocytosis by macrophages. Curr Biol 2001; 11:195-199.
17. Hoffmann PR, deCathelineau AM, Ogden CA et al. Phosphatidylserine (PS) induces PS receptor-mediated macropinocytosis and promotes clearance of apoptotic cells. J Cell Biol 2001; 155:649-659.

18. Tosello-Trampont AC, Nakada-Tsukui K, Ravichandran KS. Engulfment of apoptotic cells is negatively regulated by Rho-mediated signaling. J Biol Chem 2003; 278:49911-49919.
19. Albert ML, Kim JI, Birge RB. alphavbeta5 integrin recruits the CrkII-Dock180-rac1 complex for phagocytosis of apoptotic cells. Nat Cell Biol 2000; 2:899-905.
20. Brugnera E, Haney L, Grimsley C, et al. Unconventional Rac-GEF activity is mediated through the Dock180-ELMO complex. Nat Cell Biol 2002; 4:574-582.
21. Caron E. Rac and roll over the corpses. Curr Biol 2000; 10:R489-491.
22. Reddien PW, Horvitz HR. CED-2/CrkII and CED-10/Rac control phagocytosis and cell migration in Caenorhabditis elegans. Nat Cell Biol 2000; 2:131-136.
23. Gumienny TL, Brugnera E, Tosello-Trampont AC et al. CED-12/ELMO, a novel member of the CrkII/Dock180/Rac pathway, is required for phagocytosis and cell migration. Cell 2001; 107:27-41.
24. Zhou Z, Caron E, Hartwieg E et al. The C. elegans PH domain protein CED-12 regulates cytoskeletal reorganization via a Rho/Rac GTPase signaling pathway. Dev Cell 2001; 1:477-489.
25. Seastone DJ, Lee E, Bush J et al. Overexpression of a novel rho family GTPase, RacC, induces unusual actin-based structures and positively affects phagocytosis in Dictyostelium discoideum. Mol Biol Cell 1998; 9:2891-2904.
26. Pearson AM, Baksa K, Ramet M et al. Identification of cytoskeletal regulatory proteins required for efficient phagocytosis in Drosophila. Microbes Infect 2003; 5:815-824.
27. Kunda P, Craig G, Dominguez V et al. Abi, Sra1, and Kette control the stability and localization of SCAR/WAVE to regulate the formation of actin-based protrusions. Curr Biol 2003; 13:1867-1875.
28. Eden S, Rohatgi R, Podtelejnikov AV et al. Mechanism of regulation of WAVE1-induced actin nucleation by Rac1 and Nck. Nature 2002; 418:790-793.
29. Rivero F, Somesh BP. Signal transduction pathways regulated by Rho GTPases in Dictyostelium. J Muscle Res Cell Motil 2002; 23:737-749.
30. Chen W, Lim HH, Lim L. The CDC42 homologue from Caenorhabditis elegans. Complementation of yeast mutation. J Biol Chem 1993; 268:13280-13285.
31. Castellano F, Montcourrier P, Guillemot JC et al. Inducible recruitment of Cdc42 or WASP to a cell-surface receptor triggers actin polymerization and filopodium formation. Curr Biol 1999; 9:351-360.
32. Allen LA, Aderem A. Molecular definition of distinct cytoskeletal structures involved in complement- and Fc receptor-mediated phagocytosis in macrophages. J Exp Med 1996; 184:627-637.
33. Olazabal IM, Caron E, May RC et al. Rho-kinase and myosin-II control phagocytic cup formation during CR, but not FcgammaR, phagocytosis. Curr Biol 2002; 12:1413-1418.
34. Scott G, Leopardi S, Parker L et al. The proteinase-activated receptor-2 mediates phagocytosis in a Rho-dependent manner in human keratinocytes. J Invest Dermatol 2003; 121:529-541.
35. Morehead J, Coppens I, Andrews NW. Opsonization modulates Rac-1 activation during cell entry by Leishmania amazonensis. Infect Immun 2002; 70:4571-4580.
36. Eugene E, Hoffmann I, Pujol C et al. Microvilli-like structures are associated with the internalization of virulent capsulated Neisseria meningitidis into vascular endothelial cells. J Cell Sci 2002; 115:1231-1241.
37. Cossart P, Pizarro-Cerda J, Lecuit M. Invasion of mammalian cells by Listeria monocytogenes: functional mimicry to subvert cellular functions. Trends Cell Biol 2003; 13:23-31.
38. Mecsas J, Raupach B, Falkow S. The Yersinia Yops inhibit invasion of Listeria, Shigella and Edwardsiella but not Salmonella into epithelial cells. Mol Microbiol 1998; 28:1269-1281.
39. Bierne H, Gouin E, Roux P et al. A role for cofilin and LIM kinase in Listeria induced phagocytosis. J Cell Biol 2001; 155:101-112.
40. Mounier J, Laurent V, Hall A et al. Rho family GTPases control entry of Shigella flexneri into epithelial cells but not intracellular motility. J Cell Sci 1999; 112:2069-2080.
41. Tran Van Nhieu G, Caron E, Hall A et al. IpaC induces actin polymerisation and filopodia formation during Shigella entry into epithelial cells. EMBO J 1999; 18:3249-3262.
42. Uchiya K, Tobe T, Komatsu K et al. Identification of a novel virulence gene, virA, on the large plasmid of Shigella, involved in invasion and intercellular spreading. Mol Microbiol 1995; 17:241-250.

43. Yoshida S, Katayama E, Kuwae A et al. Shigella deliver an effector protein to trigger host microtubule destabilization, which promotes Rac1 activity and efficient bacterial internalization. EMBO J 2002; 21:2923-2935.
44. Dumenil G, Sansonetti P, Tran Van Nhieu G. Src tyrosine kinase activity down-regulates Rho-dependent responses during Shigella entry into epithelial cells and stress fibre formation. J Cell Sci 2000; 113:71-80.
45. Wiedemann A, Linder S, Grassl G et al. Yersinia enterocolitica invasin triggers phagocytosis via beta1 integrins, CDC42Hs and WASp in macrophages. Cell Microbiol 2001; 3:693-702.
46. Guzman-Verri C, Chaves-Olarte E, von Eichel-Streiber C et al. GTPases of the Rho subfamily are required for Brucella abortus internalization in nonprofessional phagocytes: direct activation of Cdc42. J Biol Chem 2001; 276:44435-44443.
47. Alrutz MA, Srivastava A, Wong KW et al. Efficient uptake of Yersinia pseudotuberculosis via integrin receptors involves a Rac1-Arp 2/3 pathway that bypasses N-WASP function. Mol Microbiol 2001; 42:689-703.
48. McGee K, Zettl M, Way M et al. A role for N-WASP in invasin-promoted internalisation. FEBS Lett 2001; 509:59-65.
49. Aepfelbacher M, Trasak C, Wiedemann A et al. Rho-GTP binding proteins in Yersinia target cell interaction. Adv Exp Med Biol 2003; 529:65-72.
50. Boquet P, Lemichez E. Bacterial virulence factors targeting Rho GTPases: parasitism or symbiosis? Trends Cell Biol 2003; 13:238-246.
51. Zumbihl R, Aepfelbacher M, Andor A et al. The cytotoxin YopT of Yersinia enterocolitica induces modification and cellular redistribution of the small GTP-binding protein RhoA. J Biol Chem 1999; 274:29289-29293.
52. Shao F, Merritt PM, Bao Z et al. A Yersinia effector and a Pseudomonas avirulence protein define a family of cysteine proteases functioning in bacterial pathogenesis. Cell 2002; 109:575-588.
53. Shao F, Vacratsis PO, Bao Z et al. Biochemical characterization of the Yersinia YopT protease: cleavage site and recognition elements in Rho GTPases. Proc Natl Acad Sci USA 2003; 100:904-909.
54. Aepfelbacher M, Trasak C, Wilharm G et al. Characterization of YopT effects on Rho GTPases in Yersinia enterocolitica-infected cells. J Biol Chem 2003; 278:33217-33223.
55. Von Pawel-Rammingen U, Telepnev MV, Schmidt G et al. GAP activity of the Yersinia YopE cytotoxin specifically targets the Rho pathway: a mechanism for disruption of actin microfilament structure. Mol Microbiol 2000; 36:737-748.
56. Kazmierczak BI, Engel JN. Pseudomonas aeruginosa ExoT acts in vivo as a GTPase-activating protein for RhoA, Rac1, and Cdc42. Infect Immun 2002; 70:2198-2205.
57. Goehring UM, Schmidt G, Pederson KJ et al. The N-terminal domain of Pseudomonas aeruginosa exoenzyme S is a GTPase-activating protein for Rho GTPases. J Biol Chem 1999; 274(51):36369-36372.
58. Gruenheid S, DeVinney R, Bladt F et al. Enteropathogenic E. coli Tir binds Nck to initiate actin pedestal formation in host cells. Nat Cell Biol 2001; 3:856-859.
59. Ben-Ami G, Ozeri V, Hanski E et al. Agents that inhibit Rho, Rac, and Cdc42 do not block formation of actin pedestals in HeLa cells infected with enteropathogenic Escherichia coli. Infect Immun 1998; 66:1755-1758.
60. Moreno SN, Docampo R. Calcium regulation in protozoan parasites. Curr Opin Microbiol 2003; 6:359-364.
61. Ghosh SK, Samuelson J. Involvement of p21racA, phosphoinositide 3-kinase, and vacuolar ATPase in phagocytosis of bacteria and erythrocytes by Entamoeba histolytica: suggestive evidence for coincidental evolution of amebic invasiveness. Infect Immun 1997; 65:4243-4249.
62. May RC, Caron E, Hall A et al. Involvement of the Arp2/3 complex in phagocytosis mediated by FcgammaR or CR3. Nat Cell Biol 2000; 2:246-248.
63. Insall R, Muller-Taubenberger A, Machesky L et al. Dynamics of the Dictyostelium Arp2/3 complex in endocytosis, cytokinesis, and chemotaxis. Cell Motil Cytoskeleton 2001; 50:115-128.
64. Olazabal IM, Machesky LM. Abp1p and cortactin, new "hand-holds" for actin. J Cell Biol 2001; 154:679-682.
65. Caron E. Regulation of Wiskott-Aldrich syndrome protein and related molecules. Curr Opin Cell Biol 2002; 14:82-87.

66. Leverrier Y, Lorenzi R, Blundell MP et al. Cutting edge: the Wiskott-Aldrich syndrome protein is required for efficient phagocytosis of apoptotic cells. J Immunol 2001; 166:4831-4834.

67. Lorenzi R, Brickell PM, Katz DR et al. Wiskott-Aldrich syndrome protein is necessary for efficient IgG-mediated phagocytosis. Blood 2000; 95:2943-2946.

68. Popoff MR, Chaves-Olarte E, Lemichez E et al. Ras, Rap, and Rac small GTP-binding proteins are targets for Clostridium sordellii lethal toxin glucosylation. J Biol Chem 1996; 271:10217-10224.

69. Chaves-Olarte E, Low P, Freer E et al. A novel cytotoxin from Clostridium difficile serogroup F is a functional hybrid between two other large clostridial cytotoxins. J Biol Chem 1999; 274:11046-11052.

70. Chaves-Olarte E, Freer E, Parra A et al. R-Ras glucosylation and transient RhoA activation determine the cytopathic effect produced by toxin B variants from toxin A-negative strains of Clostridium difficile. J Biol Chem 2003; 278:7956-7963.

71. Ehlers MR. CR3: a general purpose adhesion-recognition receptor essential for innate immunity. Microbes Infect 2000; 2:289-294.

72. Caron E, Self AJ, Hall A. The GTPase Rap1 controls functional activation of macrophage integrin alphaMbeta2 by LPS and other inflammatory mediators. Curr Biol 2000; 10:974-978.

73. Blom M, Tool AT, Roos D et al. Priming of human eosinophils by platelet-activating factor enhances the number of cells able to bind and respond to opsonized particles. J Immunol 1992; 149:3672-3677.

74. Kinbara K, Goldfinger LE, Hansen M et al. Ras GTPases: integrins' friends or foes? Nat Rev Mol Cell Biol 2003; 4:767-776.

75. Ravetch JV, Kinet JP. Fc receptors. Annu Rev Immunol 1991; 9:457-492.

76. Caron E. Cellular functions of the Rap1 GTP-binding protein: a pattern emerges. J Cell Sci 2003; 116:435-440.

77. Albert ML, Pearce SF, Francisco LM et al. Immature dendritic cells phagocytose apoptotic cells via alphavbeta5 and CD36, and cross-present antigens to cytotoxic T lymphocytes. J Exp Med 1998; 188:1359-1368.

78. Savill J, Dransfield I, Hogg N et al. Vitronectin receptor-mediated phagocytosis of cells undergoing apoptosis. Nature 1990; 343:170-173.

79. Finnemann SC, Rodriguez-Boulan E. Macrophage and retinal pigment epithelium phagocytosis: apoptotic cells and photoreceptors compete for alphavbeta3 and alphavbeta5 integrins, and protein kinase C regulates alphavbeta5 binding and cytoskeletal linkage. J Exp Med 1999; 190:861-874.

80. Robbins SM, Suttorp VV, Weeks G et al. A ras-related gene from the lower eukaryote Dictyostelium that is highly conserved relative to the human rap genes. Nucleic Acids Res 1990; 18:5265-5269.

81. Seastone DJ, Zhang L, Buczynski G et al. The small Mr Ras-like GTPase Rap1 and the phospholipase C pathway act to regulate phagocytosis in Dictyostelium discoideum. Mol Biol Cell 1999; 10:393-406.

82. Fey P, Stephens S, Titus MA et al. SadA, a novel adhesion receptor in Dictyostelium. J Cell Biol 2002; 159:1109-1119.

83. Yuan A, Siu CH, Chia CP. Calcium requirement for efficient phagocytosis by Dictyostelium discoideum. Cell Calcium 2001; 29:229-238.

84. Cornillon S, Pech E, Benghezal M et al. Phg1p is a nine-transmembrane protein superfamily member involved in Dictyostelium adhesion and phagocytosis. J Biol Chem 2000; 275:34287-34292.

85. Kitayama H, Sugimoto Y, Matsuzaki T et al. A ras-related gene with transformation suppressor activity. Cell 1989; 56:77-84.

86. Gulli MP, Peter M. Temporal and spatial regulation of Rho-type guanine-nucleotide exchange factors: the yeast perspective. Genes Dev 2001; 15:365-379.

87. Park HO, Kang PJ, Rachfal AW. Localization of the Rsr1/Bud1 GTPase involved in selection of a proper growth site in yeast. J Biol Chem 2002; 277:26721-26724.

88. Chubb JR, Wilkins A, Thomas GM et al. The Dictyostelium RasS protein is required for macropinocytosis, phagocytosis and the control of cell movement. J Cell Sci 2000; 113:709-719.

89. Self AJ, Caron E, Paterson HF et al. Analysis of R-Ras signalling pathways. J Cell Sci 2001; 114:1357-1366.

90. Zhang Z, Vuori K, Wang H et al. Integrin activation by R-ras. Cell 1996; 85:61-69.

91. Keely PJ, Rusyn EV, Cox AD et al. R-Ras signals through specific integrin alpha cytoplasmic domains to promote migration and invasion of breast epithelial cells. J Cell Biol 1999; 145:1077-1088.

92. Kinashi T, Katagiri K, Watanabe S et al. Distinct mechanisms of alpha 5beta 1 integrin activation by Ha-Ras and R-Ras. J Biol Chem 2000; 275:22590-22596.

93. Cannon GJ, Swanson JA. The macrophage capacity for phagocytosis. J Cell Sci 1992; 101:907-913.

94. Greenberg S, Grinstein S. Phagocytosis and innate immunity. Curr Opin Immunol 2002; 14:136-145.

95. Lennartz MR, Yuen AF, Masi SM et al. Phospholipase A2 inhibition results in sequestration of plasma membrane into electronlucent vesicles during IgG-mediated phagocytosis. J Cell Sci 1997; 110:2041-2052.

96. Desjardins M. ER-mediated phagocytosis: a new membrane for new functions. Nat Rev Immunol 2003; 3:280-291.

97. Garin J, Diez R, Kieffer S et al. The phagosome proteome: insight into phagosome functions. J Cell Biol 2001; 152:165-180.

98. Roberts RL, Barbieri MA, Ullrich J et al. Dynamics of rab5 activation in endocytosis and phagocytosis. J Leukoc Biol 2000; 68:627-632.

99. Vieira OV, Bucci C, Harrison RE et al. Modulation of Rab5 and Rab7 recruitment to phagosomes by phosphatidylinositol 3-kinase. Mol Cell Biol 2003; 23:2501-2514.

100. Harris E, Cardelli J. RabD, a Dictyostelium Rab14-related GTPase, regulates phagocytosis and homotypic phagosome and lysosome fusion. J Cell Sci 2002; 115:3703-3713.

101. Harris E, Yoshida K, Cardelli J et al. Rab11-like GTPase associates with and regulates the structure and function of the contractile vacuole system in Dictyostelium. J Cell Sci 2001; 114:3035-3045.

102. Cox D, Lee DJ, Dale BM et al. A Rab11-containing rapidly recycling compartment in macrophages that promotes phagocytosis. Proc Natl Acad Sci USA 2000; 97:680-685.

103. Donaldson JG. Multiple roles for Arf6: sorting, structuring, and signaling at the plasma membrane. J Biol Chem 2003; 278:41573-41576.

104. Zhang Q, Cox D, Tseng CC et al. A requirement for ARF6 in Fcgamma receptor-mediated phagocytosis in macrophages. J Biol Chem 1998; 273:19977-19981.

105. Allen LA, Yang C, Pessin JE. Rate and extent of phagocytosis in macrophages lacking vamp3. J Leukoc Biol 2002; 72:217-221.

106. Knaus UG, Heyworth PG, Evans T et al. Regulation of phagocyte oxygen radical production by the GTP-binding protein Rac 2. Science 1991; 254:1512-1515.

107. Abo A, Pick E, Hall A et al. Activation of the NADPH oxidase involves the small GTP-binding protein p21rac1. Nature 1991; 353:668-670.

Regulation of Phagocytosis by FcγRIIb and Phosphatases

Susheela Tridandapani and Clark L. Anderson

Abstract

Phagocytosis of immune-complexes is a dynamic process that is accompanied by the generation of inflammatory/tissue damaging products. Recent advances in the field indicate that this process is subject to regulation by inhibitory Fcγ receptors and intracellular phosphatases, including the inositol phosphatases SHIP-1, SHIP-2 and PTEN, and the protein tyrosine phosphatase SHP-1. This chapter will describe the role of the inhibitory Fc receptor, FcγRIIb, and the phosphatases in modulating the signaling events leading to phagocytosis and the accompanying inflammation.

Macrophages play an important role in the adaptive immune response by phagocytosing infectious particles via FcγR and complement receptors. This process is accompanied by the generation of reactive oxygen and nitrogen radicals and inflammatory cytokines in an effort to eliminate the infection. However, when produced in excess, these products can cause host tissue damage. The generation of these harmful byproducts mandates that the phagocytic process, specifically the production of inflammatory mediators, be subject to a tight regulation both in terms of the magnitude of the response, and in being contained to the locale in which the response takes place. Phagocytosis and in fact immune responses in general must be regulated, and must return to basal level after the infectious agent is eliminated. For many years the study of FcγR-mediated function of macrophages focused on molecular events leading to the activation process. Only recently have experiments revealed that FcγR-mediated activation processes are regulated; i.e., they are not simply turned on or off, but are subject to homeostatic control, resulting from the action of certain inhibitory receptors and enzymes that act alongside the activation cascade and serve to temper the biologic response.

Signaling Mechanisms Initiated by Macrophage ITAM-Associated (Activating) FcγR

Macrophages express three classes of FcγR: FcγRI, FcγRII and FcγRIII.[1] Although the expression of FcγRI and III is common to both mouse and human macrophages, the expression of FcγRII is different between the two species.[2,3] Thus, human macrophages express two functionally different FcγRII, -IIa and IIb—the products of two separate genes. In

Molecular Mechanisms of Phagocytosis, edited by Carlos Rosales. ©2005 Eurekah.com and Springer Science+Business Media.

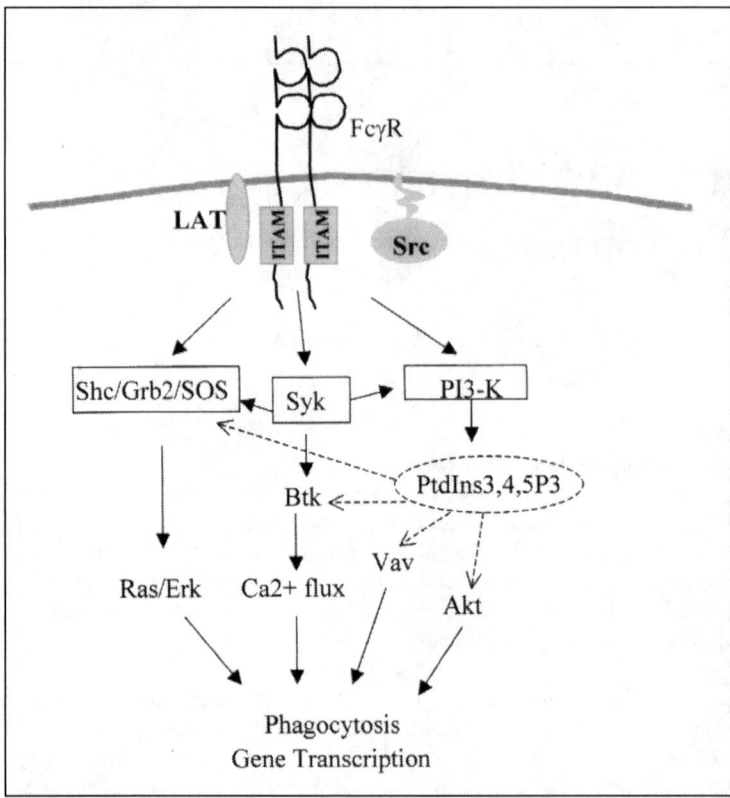

Figure 1. Signal transduction by ITAM-FcγR. Clustering of FcγR by immune-complexes results in the phosphorylation of the tyrosine-based activation motifs (ITAM) by Src kinases. Phosphorylated tyrosines of the receptor ITAMs, as well as those of the adapter LAT serve to recruit SH2 domain-containing proteins, leading to the activation of multiple signaling pathways, culminating in phagocytosis of the immune-complex and the generation of inflammatory mediators.

contrast mouse macrophages express FcγRIIb but lack the gene for FcγRIIa. FcγRI, III and IIa are activating receptors, associated with tyrosine-based activation motifs (ITAM). Whereas the ITAM of FcγRIIa is present within its cytoplasmic tail,[4] FcγRI and FcγRIIIa signal via the ITAMs present in the low molecular weight γ-subunit homodimers associated with these receptors.[5] These ITAM-FcγR, when clustered, transduce signals in a manner similar to the B cell antigen receptor (BCR) and the T cell receptor (TCR).[6-8] The earliest signaling events that occur upon FcγR clustering include activation of the receptor-associated Src kinases that phosphorylate the ITAM[9-12] (Fig. 1). The phosphorylated ITAM then serves as the docking site for Src Homology 2 (SH2) domain-containing cytoplasmic enzymes and adapter-enzyme complexes. Thus, the tyrosine kinase Syk is recruited to the phosphorylated ITAM, following which Syk becomes activated and phosphorylates a number of cytoplasmic signaling proteins.[13-15] The lipid raft-associated adapter protein LAT is also constitutively associated with FcγR and becomes phosphorylated upon receptor clustering.[16] The phosphorylated receptor ITAMs, as well as the phosphorylated tyrosines of LAT provide multiple docking sites, recruiting and activating key SH2-domain-containing proteins. Such proteins include PLCγ,[17] which hydrolyzes inositol phospholipids to generate intracellular mediators; the Ras activating GEF (guanine exchange factor) Sos, recruited through adapter proteins Shc and Grb2;[18-20] and PtdIns3-Kinase, recruited through the p85 adapter protein.[21-23] Recruitment of these effectors to the plasma membrane delivers them to the proximity of kinases that phosphorylate and activate these effectors, and also facilitates access to their protein and lipid substrates. The signaling pathways thus initiated result in a variety of functional outcomes. In the case of macrophages these outcomes include phagocytosis of IgG-opsonized particles, the generation of reactive oxygen species and the production of inflammatory cytokines.

The molecular details of FcγR-mediated phagocytosis and the production of inflammatory cytokines have been well described.[24,25] The phagocytic process is initiated when an IgG-coated particle encounters a macrophage and engages FcγR. Macrophage FcγR thus clustered then initiate intracellular signaling pathways that direct pseudopod extension around the particle and its subsequent engulfment by the macrophage. Several obligatory molecular events have been identified, including the recruitment of PtdIns3-Kinase to the plasma membrane and the subsequent phosphorylation of its lipid substrates to generate 3-phosphoinositides,[26,27] particularly PtdIns3,4,5P$_3$, which is necessary for the activation of Vav. Vav is a guanine nucleotide exchange factor (GEF) for the low molecular weight GTPases of the Rho family that promote actin polymerization and cytoskeletal rearrangements necessary for the phagocytic process. Likewise, the molecular events leading to the generation of cytokines following FcγR clustering have been extensively investigated. Inflammatory cytokines such as IL-1, IL-8 and TNF-α are generated in response to FcγR clustering and are dependent upon the activation of transcription factors such as NFAT and NFκB.[28]

Paradoxically, the ITAM-FcγR not only recruit and activate positive signaling enzymes described above, but they also associate with inhibitory enzymes such as the inositol phosphatases SHIP-1, SHIP-2, and the protein tyrosine phosphatase SHP-1. Clustering of ITAM-FcγR on human monocytes induces phosphorylation of SHIP-1 and SHIP-2 on tyrosine residues.[29-31] The activation of SHIPs by the ITAM-FcγR results in a down regulation of both phagocytosis and cytokine gene expression.[32,33] Likewise, SHP-1 phosphatase activity is turned on upon FcγRIIa clustering and results in the downregulation of phagocytosis and NFκB-dependent gene transcription.[34] These observations suggest that the activation process initiated by ITAM-FcγR clustering is subject simultaneously to the inhibitory influence of phosphatases, such that the resultant biologic response is tempered. The role of each of these phosphatases will be dicussed in greater detail in the following sections.

Negative Regulation by the ITIM-Bearing FcγRIIb

In contrast to the ITAM-FcγR described above, FcγRIIb is a receptor that serves solely to inhibit the activation processes initiated by the ITAM-bearing immunoreceptors. Human hematopoetic cells express two forms of this receptor, FcγRIIb1 and FcγRIIb2, which result from the splicing of a 19 amino acid insertion into the cytoplasmic tail of FcγRIIb1. The two isoforms exhibit differential expression patterns, with FcγRIIb1 predominating in B cells and FcγRIIb2 in monocytes/macrophages. Until recently, the lack of specific antibodies for human FcγRIIb handicapped the identification and functional characterization of this receptor in human monocytes where multiple FcγR are expressed. However, although study of the function of FcγRIIb in macrophages has lagged, great advances have been made in the understanding of FcγRIIb1 function in B cells and mast cells. The function of FcγRIIb as an inhibitory receptor was first established in B cells where cellular activation by IgG anti-IgM crosslinking was abrogated when intact IgG was used, but not if (Fab')2 fragments of IgG antibody were used.[35] It was concluded that the intact antibody co-clustered BCR with FcγRIIb, the only FcγR expressed in B cells. That the inhibitory potential of FcγRIIb resides in a 13 amino acid sequence

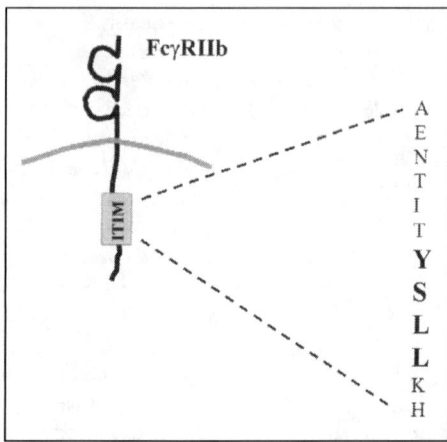

Figure 2. FcγRIIb. The cytoplasmic tail of FcγRIIb contains a tyrosine based inhibitory motif (ITIM), that becomes phosphorylated by membrane-associated Src kinases, leading to the recruitment of SH2 domain-containing phosphatases.

that contains an ITIM was elegantly demonstrated by mutational analysis of the cytoplasmic tail of the receptor[36-38] (Fig. 2). Furthermore, the in vivo role of FcγRIIb in the regulation of B cell, mast cell, and macrophage function has been assessed in genetically altered mice that are deficient in FcγRIIb expression. Thus, FcγRIIb knockout mice display elevated levels of serum IgG in response to antigenic challenge suggesting that this receptor is required for negative feedback regulation of antibody production.[39] Likewise, the mast cells of these mice also display augmented IgG-mediated degranulation, thereby supporting a regulatory role for FcγRIIb in mast cell reactivity to IgG-immune complexes. The inhibitory role of FcγRIIb phagocytosis of IgG-opsonized particles has been demonstrated by studies showing enhanced phagocytic capacity of FcγRIIb-deficient macrophages compared with wild-type macrophages.[40]

The first studies to demonstrate that FcγRIIb inhibits phagocytosis of IgG-opsonized particles were done in a transfected COS-1 fibroblast model.[41] In these studies it was shown that COS-1 fibroblasts transfected with ITAM containing FcγR were able to efficiently phagocytose IgG-coated particles. However, when the cells were also co-transfected to express FcγRIIb there was a dramatic decrease in phagocytic efficiency. These findings were later confirmed by studies showing that FcγRII-deficient murine macrophages display enhanced phagocytic ability. These observations suggested that when a monocyte/macrophage encounters an immune-complex, both the activating FcγR and inhibitory FcγR are clustered so that the magnitude of the ensuing phagocytic response is dictated by the ratio of the activating to inhibitory FcγR. Accordingly, in human monocytic cells treated with the anti-inflammatory cytokine IL-4 expression of FcγRIIb was upregulated, accompanied by decrease in phagocytic efficiency.[42,43]

The mechanism by which FcγRIIb mediates its inhibitory effects was proposed to involve the recruitment SH2 domain-containing enzymes to the phosphorylated ITIM. Using as a bait a synthetic phosphopeptide corresponding to the phosphorylated ITIM of FcγRIIb, enzymes such as the protein tyrosine phosphatase SHP-1 and the inositol phosphatase SHIP-1 were initially identified as potential candidates.[44-46] Further analyses revealed that SHIP-1 and SHP-1 do not serve redundant roles, but rather that SHIP-1 is the effector molecule of FcγRIIb-mediated inhibition, whereas SHP-1 works in concert with other ITIM-bearing receptors such as the killer inhibitory receptor (KIR) expressed on NK cells.[47-49]

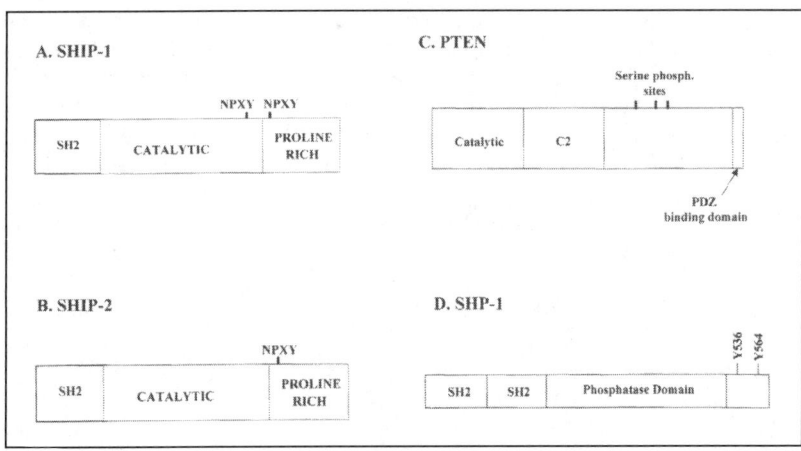

Figure 3. Phosphatases involved in FcγR signaling. A, B) SHIP-1 and SHIP-2 are inositol 5-phosphatases. C) PTEN is an inositol 3-phosphatase with some protein tyrosine phosphatase activity. D) SHP-1 is a protein tyrosine phosphatase.

The SH2 Domain Containing Inositol 5' Phosphatase SHIP-1

The hematopoietic cell-specific inositol phosphatase SHIP-1 is a multi-domain cytosolic protein, containing an inositol 5' phosphatase domain and several protein interaction domains[50-52] (Fig. 3). The interaction domains of SHIP-1 include: (a) a SH2 domain that associates with high affinity to the phosphorylated ITIM of FcγRIIb,[45,53,54] (b) a proline-rich domain that is constitutively associated with the SH3 domains of the Ras adapter protein Grb2,[55] and 3) two sites of tyrosine phosphorylation set in a motif that preferentially binds proteins containing phosphotyrosine binding (PTB) domains, such as the Ras adapter Shc,[51,56,57] and the RasGAP-binding protein Dok.[58] SHIP-1 hydrolyzes PtdIns-3,4,5P$_3$, a product of PtdIns3-Kinase, which is required for the activation of several key enzymes that contain a plextrin homology (PH) domain, such as the tyrosine kinase Btk,[59] involved in intracellular calcium mobilization, Akt,[60-62] an enzyme involved in protecting cells from apoptosis, and Vav.[63] Although the enzyme activity of SHIP is constitutively turned on, SHIP must localize to the plasma membrane to access its lipid substrates. Thus, in resting cells SHIP is in the cytoplasm and translocates to the membrane upon activation of the cell. Membrane association of SHIP is mediated by the association of its SH2 domain with tyrosine phosphorylated receptors either directly, or through the adapter molecule Shc. The C-terminal region of SHIP appears to further enhance membrane-association, and the stability of SHIP at the membrane.[64-66]

The inhibitory role of SHIP-1 on BCR-mediated activation of B cells, FcγRI-mediated activation of mast cells, and FcγR-mediated activation of macrophages is well established both in vitro using cell lines and in vivo using SHIP-1 knockout mice.[67] Thus, the presence of SHIP-1 down regulates BCR-mediated calcium flux,[45] MAPKinase activation,[55] and Akt activation,[60,61] and FcγRI-mediated mast-cell degranulation.[68] Likewise, SHIP has also been shown to dampen the activation events initiated by growth factor and cytokine receptors.[67]

The role of SHIP in phagocytosis is actively being investigated by several groups. The initial studies to examine the influence of SHIP on phagocytosis used macrophages derived from SHIP knockout mice and their wild type littermates, and found that deletion of SHIP resulted in enhanced phagocytic efficiency.[69] Other studies soon followed, analyzing the molecular details of SHIP activation by macrophage FcγR. These studies established that SHIP is not only capable of associating with phosphorylated ITIMs, but is also able to associate with the phosphorylated ITAMs of the activating FcγR both directly as well as through the adapter molecule Shc.[32,33] In addition to downregulating phagocytosis, SHIP is also capable of dampening FcγR-induced activation of NFκB in monocytic cells.[33] Together these studies suggest that SHIP serves to temper FcγR-mediated phagocytosis and the accompanying inflammatory cytokine gene expression working through both activating and inhibiting FcγR.

Despite the close attention that SHIP has received in the last few years, there remain several unanswered questions with regard to the role of SHIP in macrophage FcγR function. For example, it is well known that SHIP can influence signaling pathways not only by its ability to hydrolyze PtdIns3,4,5P3, but also by its ability to interact with key signaling molecules via the SHIP interaction domains. In B cells, SHIP appears to downregulate the Ras signaling pathway by at least two mechanisms that do not involve SHIP catalytic function. In the first mechanism proposed SHIP competitively inhibits the association of Shc with the Grb2/Sos complex, preventing Grb2/Sos membrane translocation and the subsequent activation of Ras.[70,71] Second, SHIP is reported to associate with p62dok, a RasGAP-binding protein, and promote the inactivation of Ras.[58] At present, it is unclear whether SHIP influences macrophage FcγR function by solely its enzyme activity, hydrolyzing PtdIns3,4,5P3, or whether there are additional mechanisms involving the protein interaction domains of SHIP.

The Inositol Phosphatase SHIP-2

SHIP-2 is a SH2 domain-containing inositol 5' phosphatase that has a high level homology to SHIP-1 in its catalytic region (Fig. 3).[72,73] The molecules are largely divergent in the C-terminal region consisting of a proline-rich domain that associates with unique SH3 domain-containing proteins.[74] Thus, while the proline-rich domain of SHIP-1 associates specifically with Grb2, the proline-rich domain of SHIP-2 associates with other SH3 domain containing proteins such as Abl. In addition, while SHIP-1 has two tyrosine residues in the C-terminus that conform to an NPXY motif shown to bind PTB (phosphotyrosine binding) domains upon phosphorylation, SHIP-2 has only one NPXY motif.[75] Thus, these two enzymes while enzymatically similar, likely differ in functions that are related to their protein interactions via the C-terminal region. These two molecules also differ in their expression patterns: SHIP-1 is expressed predominantly in hematopoetic cells, while SHIP-2 is much more ubiquitously expressed. Recent studies have revealed a critical role for SHIP-2 in regulating insulin receptor signaling.[76-78] Gene knockout mice that are deficient in SHIP-2 expression are hypoglycemic and do not survive past the first day. Other studies have demonstrated a role for SHIP-2 in compensating for the loss of SHIP-1 in SHIP-1-deficient B cell blasts, and mediating the inhibitory effect of FcγRIIb.[79,80]

SHIP-2 is abundantly expressed in macrophages but is present at virtually undetectable levels in human peripheral blood monocytes. However, SHIP-2 expression in PBM is dramatically upregulated upon treatment with bacterial lipopolysaccharide.[30] Upon FcγR clustering, SHIP-2 is recruited to the phosphorylated ITAM of FcγRIIa, in a manner that is dependent on the SH2 domain of SHIP-2, and serves to down regulate FcγRIIa-induced activation of Akt and NFκB-dependent gene transcription. Although a role for SHIP-2 in phagocytosis has not yet been reported, based on the findings that SHIP-2, like SHIP-1, dampens FcγR-induced signaling events such as the activation of Akt and NFκB would suggest that SHIP-2 likely also downregulates phagocytosis.

PTEN (Phosphatase and Tensin Homologue on Chromosome 10)

PTEN is a dual phosphatase that can dephosphorylate phospholipids at the 3' position, as well as tyrosine-phosphorylated proteins (Fig. 3).[81,82] The enzyme consists of a N-terminal phosphatase domain, a C2 domain and a C-terminal PDZ-binding sequence and multiple phosphorylation sites.[67] The C2 domain and the phosphatase domain have been shown to be necessary for membrane localization of PTEN.[83] In contrast, phosphorylation of PTEN is accompanied by dissociation of PTEN from the membrane and a reduction in phosphatase activity. PTEN has been reported to negatively regulate immune receptor and growth factor receptor-mediated events. Most of PTEN's biological effects are attributed to its inositol phosphatase activity. For example, PTEN is thought to serve as a tumor suppressor by hydrolyzing PtdIns3,4,5P$_3$ and thereby downregulating the activation of the Akt. The protein tyrosine phosphatase activity of PTEN has been reported to regulate the Ras/Erk pathway by dephosphorylating the Ras adapter Shc,[84] and influence adhesion and cell migration by dephosphorylating focal adhesion kinase (FAK).[85,86]

The role of PTEN in phagocytosis is largely unknown, although there is some evidence based on a transfected fibroblast model that PTEN may suppress phagocytosis through the downregulation of Rac activation.[87] The molecular mechanisms involved in the activation of PTEN by FcγR, and the functional role of PTEN's lipid phosphatase and protein phosphatase activity in phagocytosis remain to be investigated in macrophages.

The Protein Tyrosine Phosphatase SHP-1

SHP-1 is 65kD protein tyrosine phosphatase expressed predominantly in hematopoetic cells. SHP-1 contains two N-terminal SH2 domains, a phosphatase domain and two tyrosine phosphorylation sites in the C-terminal region of the protein[88-90] (Fig. 3). The phosphatase activity of SHP-1 is regulated by intramolecular interactions, such that the N-terminal SH2 domain is folded over the phosphatase domain in the inactive state.[91,92] The enzyme becomes activated when the N-terminal SH2 domain is engaged by a phosphopeptide, which allows the phosphatase domain to gain access to its substrate. Thus SHP-1 activity can be upregulated by either addition of a cognate phosphopeptide or by deletion of the N-terminal SH2 domain. In contrast, the C-terminal SH2 domain serves to recruit SHP-1 substrates but does not contribute to activation of the enzyme. Recent evidence suggests that of the two C-terminal tyrosines that become phosphorylated, modification of Y-536 results in a 4-fold increase in SHP-1 activity. In contrast, modification of Y564 did not alter SHP-1 phosphatase activity. Thus phosphorylation of SHP-1 on tyrosine 536 contributes to activation of the enzyme.[93]

The functional role of SHP-1 as a negative regulator emerged with the identification of mice with point mutations in the SHP-1 gene such that the result was the expression of a catalytically inactive splice variant in one mutant, or a complete loss of SHP-1 protein expression in the other. The animals bearing the former mutations are designated motheaten viable (Mev), and those bearing the latter mutation are designated as motheaten (Me).[94-97] Both of these mutations result in a severe phenotype, with uncontrolled expansion of myeloid cells and a shortened life span of about 5-10 weeks. Specifically, T and B cells from SHP-1 deficient animals are hyperresponsive to immune receptor stimulation.[98-102] In addition, SHP-1 has been demonstrated to negatively regulate a wide variety of growth factor receptors including the CSF-1 receptor and EGF receptor.[89] Among the immunereceptors known to invoke SHP-1 are the Killer cell Inhibitory Receptors (KIR),[49,103,104] CD22[105] and gp49B.[106,107] Although initial studies found association of SHP-1 to FcγRIIb derived peptides, more recent studies have conclusively demonstrated that FcγRIIb-mediated inhibition occurs via SHIP and not SHP-1.[47-49]

Recent studies have addressed the role of SHP-1 in FcγR-mediated activation of macrophages. Thus, Durden and colleagues found that human FcγRIIa-mediated phagocytosis was downregulated by SHP-1 in a transfected fibroblast model.[108] In this system over-expression of SHP-1 was shown to lead to dephosphorylation of Cbl and downregulation of Rac activation. Other studies demonstrated that SHP-1 downregulates NFκB-dependent gene transcription in response to FcγRIIa clustering in monocytic cells.[34] In these latter studies SHP-1 was reported to associate specifically with the phosphorylated N-terminal ITAM tyrosine of FcγRIIa. Taken together, these latter observations would imply that SHP-1 is activated in macrophages during FcγR-mediated phagocytosis, perhaps in a non-FcγRIIb-dependent manner. The molecular details of SHP-1 activation by FcγR and the precise nature of SHP-1 influence on FcγR-mediated signaling biology remains to be examined. Additional work is needed to identify the exact substrates of SHP-1 during the phagocytic process.

It is clear from the studies described above that immune-complex clearance is a highly regulated process, and that the absence of such regulation could lead to inflammation and tissue damage. However, it is not yet entirely clear why there exist so many different mechanisms of regulation. There remain many unanswered questions regarding the non-overlapping functions of the phosphatases involved. As well, these regulatory mechanisms may be separated by expression pattern of the phosphatases during differentiation stages of myeloid cells. As future work defines the exact molecular details of these regulatory mechanisms it will be interesting to see whether there might be molecules that will specifically downregulate the production of inflammatory mediators while maintaining the efficiency of engulfment and destruction of IgG-coated particles.

References

1. Anderson CL, Shen L, Eicher DM et al. Phagocytosis mediated by three distinct Fcγ receptor classes on human leukocytes. J Exp Med 1990; 171:1333-1346.
2. Ravetch JV, Bolland S. IgG Fc receptors. Annu Rev Immunol 2001; 19:275-290.
3. Daeron M. Fc receptor biology. Annu Rev Immunol 1997; 15:203-234.
4. Van Den Herik-Oudijk IE, Capel PJ, Van der Bruggen T et al. Identification of signaling motifs within human FcγRIIa and FcγRIIb isoforms. Blood 1995; 85:2202-2211.
5. Ernst LK, Duchemin A-M, Anderson CL. Association of the high affinity receptor for IgG (FcγRI) with the γ subunit of the IgE receptor. Proc Natl Acad Sci USA 1993; 90:6023-6027.
6. Weiss A. T cell antigen receptor signal transduction: A tale of tails and cytoplasmic protein-tyrosine kinases. Cell 1993; 209-212.
7. Kurosaki T. Molecular mechanisms in B cell antigen receptor signaling. Curr Opin Immunol 1997; 9:309-318.
8. Reth M. Antigen receptor tail clue. Nature 1989; 338:383-384.
9. Duchemin A-M, Anderson CL. Association of non-receptor protein tyrosine kinases with the FcγRI/γ-chain complex in monocytic cells. J Immunol 1997; 158:865-871.
10. Ghazizadeh S, Fleit HB. Tyrosine phosphorylation provides an early obligatory signal for FcγRII-mediated endocytosis in the monocytic cell line THP-1. J Immunol 1994; 152:30-41.
11. Ghazizadeh S, Bolen JB, Fleit HB. Physical and functional association of Src-related protein tyrosine kinases with FcγRII in monocytic THP-1 cells. J Biol Chem 1994; 269:8878-8884.
12. Hamada F, Aoki M, Akiyama T et al. Association of immunoglobulin FcγRII with Src-like protein-tyrosine kinase Fgr in neutrophils. Proc Natl Acad Sci USA 1993; 90:6305-6309.
13. Ghazizadeh S, Bolen JB, Fleit HB. Tyrosine phosphorylation and association of Syk with FcγRII in monocytic THP-1 cells. Biochem J 1995; 305:669-674.
14. Chacko GW, Duchemin A-M, Coggeshall KM et al. Clustering of the platelet FcγR induces noncovalent association with the tyrosine kinase p72syk. J Biol Chem 1994; 269:32435-32440.
15. Kiener PA, Rankin BM, Burkhardt AL et al. Cross-linking of Fcγ receptor I (FcγRI) and receptor II (FcγRII) on monocytic cells activates a signal transduction pathway common to both Fc receptors that involves the stimulation of p72 Syk protein tyrosine kinase. J Biol Chem 1993; 268:24442-24448.
16. Tridandapani S, Lyden TW, Smith JL et al. The adapter protein LAT enhances Fcγ receptor-mediated signal transduction in myeloid cells. J Biol Chem 2000; 275:20480-20487.

17. Liao F, Shin HS, Rhee SG. Tyrosine phosphorylation of phospholipase C-γ1 induced by cross-linking of the high-affinity or low-affinity Fc receptor for IgG in U937 cells. Proc Natl Acad Sci USA 1992; 89:3659-3663.
18. Margolis B. The GRB family of SH2 domain proteins. Prog Biophys Mol Biol 1994; 62:223-244.
19. Downward J. Control of ras activation. Cancer Surv 1996; 27:87-100:87-100.
20. Downward J. The GRB2/Sem-5 adaptor protein. FEBS Lett 1994; 338:113-117.
21. Chacko GW, Brandt JT, Coggeshall KM et al. Phosphatidylinositol 3-kinase and p72Syk noncovalently associate with the low affinity Fc gamma receptor on human platelets through an ITAM: reconstitution with synthetic phosphopeptides. J Bio Chem 1996; 271:10775-10781.
22. Gibbins J, Briddon S, Shutes A et al. The p85 subunit of phosphatidylinositol 3-kinase associates with the Fc receptor. J Biological Chem 1998; 273:34437-34443.
23. Cooney DS, Phee H, Jacob A et al. Signal transduction by human-restricted FcγRIIa involves three distinct cytoplasmic kinase families leading to phagocytosis. J Immunol 2001; 167:844-854.
24. Aderem A, Underhill DM. Mechanisms of phagocytosis in macrophages. Annu Rev Immunol 1999; 17:593-623.
25. Sánchez-Mejorada G, Rosales C. Signal transduction by immunoglobulin Fc receptors. J Leukoc Biol 1998; 63:521-533.
26. Lowry MB, Duchemin A-M, Coggeshall KM et al. Chimeric receptors composed of PI3-kinase domains and Fcγ receptor ligand-binding domains mediate phagocytosis in COS fibroblasts. J Biol Chem 1998; 273:24513-24520.
27. Cox D, Tseng CC, Bjekic G et al. A requirement for phosphatidylinositol 3-kinase in pseudopod extension. J Biol Chem 1999; 274:1240-1247.
28. Sanchez-Mejorada G, Rosales C. Fcgamma receptor-mediated mitogen-activated protein kinase activation in monocytes is independent of Ras. J Biol Chem 1998; 273:27610-27619.
29. Maresco DL, Osborne JM, Cooney D et al. The SH2-containing 5'-inositol phosphatase (SHIP) is tyrosine phosphorylated after Fcγ receptor clustering in monocytes. J Immunol 1999; 162:6458-6465.
30. Pengal RA, Ganesan LP, Fang H et al. SHIP-2 inositol phosphatase is inducibly expressed in human monocytes and serves to regulate Fcγ receptor-mediated signaling. J Biol Chem 2003; 278:22657-63.
31. Cameron AJ, Allen JM. The human high-affinity immunoglobulin G receptor activates SH2-containing inositol phosphatase (SHIP). Immunology 1999; 97:641-647.
32. Nakamura K, Malykhin A, Coggeshall KM. The Src homology 2 domain-containing inositol 5-phosphatase negatively regulates Fcγ receptor-mediated phagocytosis through immunoreceptor tyrosine-based activation motif-bearing phagocytic receptors. Blood 2002; 100:3374-3382.
33. Tridandapani S, Wang Y, Marsh CB et al. Src homology 2 domain-containing inositol polyphosphate phosphatase regulates NF-κB-mediated gene transcription by phagocytic FcγRs in human myeloid cells. J Immunol 2000; 169:4370-8.
34. Ganesan LP, Fang H, Marsh CB et al. The protein-tyrosine phosphatase SHP-1 associates with the phosphorylated immunoreceptor tyrosine-based activation motif of FcγRIIa to modulate signaling events in myeloid cells. J Biol Chem 2003; 278:35710-35717.
35. Phillips NE, Parker DC. Cross-linking of B lymphocyte Fcγ receptors and membrane immunoglobulin inhibits anti-immunoglobulin-induced blastogenesis. J Immunol 1984; 132:627-632.
36. Muta T, Kurosaki T, Misulovin Z et al. A 13-amino-acid motif in the cytoplasmic domain of Fcγ RIIB modulates B-cell receptor signalling [published erratum appears in Nature 1994; 369(6478):340]. Nature 1994; 368:70-73.
37. Amigorena S, Bonnerot C, Drake JR et al. Cytoplasmic domain heterogeneity and functions of IgG Fc receptors in B lymphocytes. Science 1992; 256:1808-1812.
38. Choquet D, Ku G, Cassard S et al. Different patterns of calcium signaling triggered through two components of the B lymphocyte antigen receptor. J Biol Chem 1994; 269:6491-6497.
39. Takai T, Ono M, Hikida M. Augmented humoral and anaphylactic responses in FcγRII-deficient mice. Nature 1996; 379:346-349.
40. Clynes R, Maizes JS, Guinamard R et al. Modulation of immune complex-induced inflammation in vivo by the coordinate expression of activation and inhibitory Fc receptors. J Exp Med 1999; 189:179-185.

41. Hunter S, Indik ZK, Kim MK et al. Inhibition of Fcγ receptor-mediated phagocytosis by a nonphagocytic Fcγ receptor. Blood 1998; 91:1762-1768.
42. Pricop L, Redecha P, Teillaud J-L et al. Differential modulation of stimulatory and inhibitory Fcγ receptors on human monocytes by Th1 and Th2 cytokines. J Immunol 2001; 166:531-537.
43. Tridandapani S, Siefker K, Teillaud J-L et al. Regulated expression and inhibitory function of FcγRIIb in human monocytic cells. J Biol Chem 2001; 277:5082-9.
44. D'Ambrosio D, Hippen KL, Minskoff SA et al. Recruitment and activation of PTP1C in negative regulation of antigen receptor signaling by FcγRIIBI. Science 1995; 268:293-296.
45. Ono M, Bolland S, Tempst P et al. Role of the inositol phosphatase SHIP in negative regulation the of immune system by the receptor FcγRIIB. Nature 1996; 383:263-266.
46. Chacko GW, Tridandapani S, Damen JE et al. Negative signaling in B lymphocytes induces tyrosine phosphorylation of the 145-kDa inositol polyphosphate 5-phosphatase, SHIP. J Immunol 1996; 157:2234-2238.
47. Ono M, Okada H, Bolland S et al. Deletion of SHIP or SHP-1 reveals two distinct pathways for inhibitory signaling. Cell 1997; 90:293-301.
48. Nadler MJS, Chen B, Anderson JS et al. Protein-tyrosine phosphatase SHP-1 is dispensable for FcγRIIB-mediated inhibition of B cell antigen receptor activation. J Bio Chem 1997; 272:20038-20043.
49. Gupta N, Scharenberg AM, Burshtyn DN et al. Negative signaling pathways of the killer cell inhibitory receptor and FcγRIIb1 require distinct phosphatases. J Exp Med 1997; 186:473-478.
50. Damen JE, Liu L, Rosten P et al. The 145-kDa protein induced to associate with Shc by multiple cytokines is an inositol tetraphosphate and phosphatidylinositol 3,4,5-triphosphate 5-phosphatase. Proc Natl Acad Sci USA 1996; 93:1689-1693.
51. Lioubin MN, Algate PA, Tsai S et al. p150 SHIP, a signal transduction molecule with inositol polyphosphate-5-phosphatase activity. Genes Dev 1996; 10:1084-1095.
52. Kavanaugh WM, Pot DA, Chin SM et al. Multiple forms of an inositol polyphosphate 5-phosphatase form signaling complexes with Shc and Grb2. Current Biology 1996; 6:438-445.
53. Tridandapani S, Kelley T, Pradhan M et al. Recruitment and phosphorylation of SHIP and Shc to the B cell Fcγ ITIM peptide motif. Mol Cell Biol 1997; 17:4305-4311.
54. Tridandapani S, Pradhan M, LaDine JR et al. Protein interactions of SHIP: association with Shc displaces SHIP from FcγRIIb in B cells. J Immunol 1998; 162:1408-1414.
55. Tridandapani S, Phee H, Shivakumar L et al. Role of Ship in FcγRIIb-mediated inhibition of Ras activation in B cells. Mol Immunol 1998; 35:1135-1146.
56. Pradhan M, Coggeshall KM. Activation-induced bi-dentate interaction of SHIP and Shc in B lymphocytes. J Cell Biochem 1997; 67:32-42.
57. Lamkin TD, Walk SF, Liu L et al. Shc interaction with Src homology 2 domain containing inositol phosphatase (SHIP) in vivo requires the Shc-phosphotyrosine binding domain and two specific phosphotyrosines on SHIP. J Bio Chem 1997; 272:10396-10401.
58. Tamir I, Stolpa JC, Helgason CD et al. The RasGAP-binding protein p62dok is a mediator of inhibitory FcγRIIB signals in B cells. Immun 2000; 12:347-358.
59. Scharenberg AM, El-Hillal O, Fruman DA et al. Phosphatidylinositol-3,4,5-trisphosphate (PtdIns-3,4,5-P$_3$) Tec kinase-dependent calcium signaling pathway: A target for SHIP-mediated inhibitory signals. EMBO J 1998; 17:1961-1972.
60. Jacob A, Cooney D, Tridandapani S et al. FcγRIIB modulation of surface immunoglobulin-induced Akt activation in murine B cells. J Biol Chem 1999; 274:13704-13710.
61. Aman MJ, Lamkin TD, Okada H et al. The inositol phosphatase SHIP inhibits Akt/PKB activation in B cells. J Biol Chem 1998; 273:33922-33928.
62. Carver DJ, Aman MJ, Ravichandran KS. SHIP inhibits Akt activation in B cells through regulation of Akt membrane localization. Blood 2000; 96:1449-1456.
63. Ma AD, Metjian A, Bagrodia S et al. Cytoskeletal reorganization by G protein-coupled receptors is dependent on phosphoinositide 3-kinase gamma, a Rac guanosine exchange factor, and Rac. Mol Cell Biol 1998; 18:4744-4751.
64. Aman MJ, Walk SF, March ME et al. Essential role for the C-terminal noncatalytic region of SHIP in FcγRIIB1-mediated inhibitory signaling. Mol Cell Biol 2000; 20:3576-3589.

65. Damen JE, Ware MD, Kalesnikoff J et al. SHIP's C-terminus is essential for its hydrolysis of PIP3 and inhibition of mast cell degranulation. Blood 2001; 97:1343-1351.

66. Baran CP, Tridandapani S, Helgason CD et al. The inositol 5'-phosphatase SHIP-1 and the Src kinase Lyn negatively regulate macrophage colony-stimulating factor-induced Akt activity. J Biol Chem 2003; 278:38628-38636.

67. Krystal G. Lipid phosphatases in the immune system. Immunology 2000; 12:397-403.

68. Huber M, Helgason CD, Damen JE et al. The src homology 2-containing inositol phosphatase (SHIP) is the gatekeeper of mast cell degranulation. Proc Natl Acad Sci 1998; 95:11330-11335.

69. Cox D, Dale BM, Kishiwada M et al. A regulatory role for Src homology 2 domain-containing inositol 5'-phosphatase (SHIP) in phagocytosis mediated by Fcγ receptors and complement receptor 3 (α(M)β(2); CD11b/CD18. J Exp Med 2001; 193:61-71.

70. Tridandapani S, Chacko GW, Van Bruggen MCJ et al. Negative signaling in B cells Causes reduced Ras activity by reducing Shc-Grb2 interactions. J Immunol 1997; 158:1125-1132.

71. Tridandapani S, Kelly T, Cooney D et al. Negative signaling in B cells: SHIP Grbs Shc. Immunol Today 1997; 18:424-427.

72. Pesesse X, Moreau C, Drayer AL et al. The SH2 domain containing inositol 5-phosphatase SHIP2 displays phosphatidylinositol 3,4,5-trisphosphate and inositol 1,3,4,5-tetrakisphosphate 5-phosphatase activity. FEBS Lett 1998; 437:301-303.

73. Pesesse X, Dewaste V, De Smedt F et al. The Src homology 2 domain containing inositol 5-phosphatase SHIP2 is recruited to the epidermal growth factor (EGF) receptor and dephosphorylates phosphatidylinositol 3,4,5-trisphosphate in EGF-stimulated COS-7 cells. J Biol Chem 2001; 276:28348-28355.

74. Wisniewski D, Strife A, Swendeman S et al. A novel SH2-containing phosphatidylinositol 3,4,5-trisphosphate 5-phosphatase (SHIP2) is constitutively tyrosine phosphorylated and associated with src homologous and collagen gene (SHC) in chronic myelogenous leukemia progenitor cells. Blood 1999; 93:2707-2720.

75. Erneux C, Govaerts C, Communi D et al. The diversity and possible functions of the inositol polyphosphate 5-phosphatases. Biochim Biophys Acta 1998; 1436:185-199.

76. Blero D, De Smedt F, Pesesse X et al. The SH2 domain containing inositol 5-phosphatase SHIP2 controls phosphatidylinositol 3,4,5-trisphosphate levels in CHO-IR cells stimulated by insulin. Biochem Biophys Res Commun 2001; 282:839-843.

77. Wada T, Sasaoka T, Funaki M et al. Overexpression of SH2-containing inositol phosphatase 2 results in negative regulation of insulin-induced metabolic actions in 3T3-L1 adipocytes via its 5'-phosphatase catalytic activity. Mol Cell Biol 2001; 21:1633-1646.

78. Clement S, Krause U, Desmedt F et al. The lipid phosphatase SHIP2 controls insulin sensitivity. Nature 2001; 409:92-97.

79. Brauweiler A, Tamir I, Marschner S et al. Partially distinct molecular mechanisms mediate inhibitory FcgammaRIIB signaling in resting and activated B cells. J Immunol 2001; 167:204-211.

80. Muraille E, Pesesse X, Kuntz C et al. Distribution of the src-homology-2-domain-containing inositol 5-phosphatase SHIP-2 in both non-haemopoietic and haemopoietic cells and possible involvement of SHIP-2 in negative signalling of B-cells. Biochem J 1999; 342.:697-705.

81. Sulis ML, Parsons R. PTEN: From pathology to biology. Trends Cell Biol 2003; 13:478-483.

82. Cantley LC, Neel BG. New insights into tumor suppression: PTEN suppresses tumor formation by restraining the phosphoinositide 3-kinase/AKT pathway. Proc Natl Acad Sci USA 1999; 96:4240-4245.

83. Das S, Dixon JE, Cho W. Membrane-binding and activation mechanism of PTEN. Proc Natl Acad Sci USA 2003; 100:7491-7496.

84. Gu J, Tamura M, Yamada KM. Tumor suppressor PTEN inhibits integrin- and growth factor-mediated mitogen-activated protein (MAP) kinase signaling pathways. J Cell Biol 1998; 143:1375-1383.

85. Gu J, Tamura M, Pankov R et al. Shc and FAK differentially regulate cell motility and directionality modulated by PTEN. J Cell Biol 1999; 146:389-403.

86. Tamura M, Gu J, Takino T et al. Tumor suppressor PTEN inhibition of cell invasion, migration, and growth: differential involvement of focal adhesion kinase and p130Cas. Cancer Res 1999; 59:442-449.

87. Kim JS, Peng X, De PK et al. PTEN controls immunoreceptor (immunoreceptor tyrosine-based activation motif) signaling and the activation of Rac. Blood 2002; 99:694-697.

88. Zhang J, Somani AK, Siminovitch KA. Roles of the SHP-1 tyrosine phosphatase in the negative regulation of cell signalling. Immunology 2000; 12:361-378.

89. Neel BG, Tonks NK. Protein tyrosine phosphatases in signal transduction. Curr Opin Cell Biol 1997; 9:193-204.

90. Neel BG. Role of phosphatases in lymphocyte activation. Curr Opin Immunol 1997; 9:405-420.

91. Pei D, Lorenz U, Klingmuller U et al. Intramolecular regulation of protein tyrosine phosphatase SH-PTP1: A new function for Src homology 2 domains. Biochemistry 1994; 33:15483-15493.

92. Pei D, Wang J, Walsh CT. Differential functions of the two Src homology 2 domains in protein tyrosine phosphatase SH-PTP1. Proc Natl Acad Sci USA 1996; 93:1141-1145.

93. Zhang Z, Shen K, Lu W et al. The role of C-terminal tyrosine phosphorylation in the regulation of SHP-1 explored via expressed protein ligation. J Biological Chem 2003; 278:4668-74

94. Shultz LD, Schweitzer PA, Rajan TV et al. Mutations at the murine motheaten locus are within the hematopoietic cell protein-tyrosine phosphatase (Hcph) gene. Cell 1993; 73:1445-1454.

95. Shultz LD, Green MC. Motheaten, an immunodeficient mutant of the mouse. II. Depressed immune competence and elevated serum immunoglobulins. J Immunol 1976; 116:936-943.

96. Green MC, Shultz LD. Motheaten, an immunodeficient mutant of the mouse. I. Genetics and pathology. J Hered 1975; 66:250-258.

97. Tsui HW, Siminovitch KA, de Souza L et al. Motheaten and viable motheaten mice have mutations in the haematopoietic cell phosphatase gene. Nat Genet 1993; 4:124-129.

98. Cornall RJ, Goodnow CC, Cyster JG. Regulation of B cell antigen receptor signaling by the Lyn/CD22/SHP1 pathway. Curr Top Microbiol Immunol 1999; 244:57-68.:57-68.

99. Doody GM, Justement LB, Delibrias CC et al. A role in B cell activation for CD22 and the protein tyrosine phosphatase SHP. Science 1995; 269:242-244.

100. Plas DR, Johnson R, Pingel JT et al. Direct regulation of ZAP-70 by SHP-1 in T cell antigen receptor signaling. Science 1996; 272:1173-1176.

101. Pani G, Fischer KD, Mlinaric-Rascan I et al. Signaling capacity of the T cell antigen receptor is negatively regulated by the PTP1C tyrosine phosphatase. J Exp Med 1996; 184:839-852.

102. Pani G, Kozlowski M, Cambier JC et al. Identification of the tyrosine phosphatase PTP1C as a B cell antigen receptor-associated protein involved in the regulation of B cell signaling. J Exp Med 1995; 181:2077-2084.

103. Berg KL, Carlberg K, Rohrschneider LR et al. The major SHP-1-binding, tyrosine-phosphorylated protein in macrophages is a member of the KIR/LIR family and an SHP-1 substrate. Oncogene 1998; 17:2535-2541.

104. Burshtyn DN, Lam AS, Weston M et al. Conserved residues amino-terminal of cytoplasmic tyrosines contribute to the SHP-1-mediated inhibitory function of killer cell Ig-like receptors. J Immunol 1999; 162:897-902.

105. Blasioli J, Goodnow CC. Lyn/CD22/SHP-1 and their importance in autoimmunity. Curr Dir Autoimmun 2002; 5:151-160.

106. Kuroiwa A, Yamashita Y, Inui M et al. Association of tyrosine phosphatases SHP-1 and SHP-2, inositol 5-phosphatase SHIP with gp49B1, and chromosomal assignment of the gene. J Biol Chem 1998; 273:1070-1074.

107. Lu-Kuo JM, Joyal DM, Austen KF et al. gp49B1 inhibits IgE-initiated mast cell activation through both immunoreceptor tyrosine-based inhibitory motifs, recruitment of src homology 2 domain-containing phosphatase-1, and suppression of early and late calcium mobilization. J Biol Chem 1999; 274:5791-5796.

108. Kant AM, De P, Peng X et al. SHP-1 regulates Fcγ receptor-mediated phagocytosis and the activation of RAC. Blood 2002; 100:1852-1859.

CHAPTER 8

Phospholipases and Phagocytosis

Michelle R. Lennartz

Introduction

Phagocytosis, the process by which particulate matter is taken up by cells, is characteristic of many cell types including the pigmented epithelium of the eye, Langerhans cells in the skin, the microglia in the brain, and Kupffer cells in the liver. Of particular interest clinically are the "professional" phagocytes: neutrophils, monocytes, and tissue macrophages. Their ability to phagocytose foreign pathogens, senescent cells, and cellular debris is essential for homeostasis. Dendritic cells should also be mentioned as they act as sentries to take up bacteria in the skin and at mucosal surfaces. However, their primary function is to present antigens and the majority of studies address this function. Thus, we will focus on macrophages and neutrophils in this review.

Phagocytosis occurs through a variety of receptors, including those for immunoglobulin G (IgG, FcγR), complement (CR1, CR3), mannose (MR), and β-glucan (dectin-1),[1] and the scavenger receptor that facilitates uptake of oxidized lipids and advanced glycation end products.[2] Although much is known about the structure of phagocytic receptors and their ligands, the signaling events that accompany ingestion are the subject of ongoing investigation with exciting advances being reported on a regular basis. Recent advances in molecular biology and imaging have provided novel tools for studying phospholipids and phospholipase products in phagocytosis. This review will summarize our current understanding of the role of phospholipases in phagocytosis. As much of our knowledge of this topic is derived from macrophage models of IgG-mediated uptake, the bulk of this review will draw on that data. Relevant studies involving other targets will be included where information is available.

Phagocytosis

The initiating event in phagocytosis is the binding of phagocyte cell surface receptors to ligands on the surface of the particle to be ingested. These ligands may be intrinsic to the target, i.e., the carbohydrate coat of yeast or pathogen-associated molecular patterns (PAMPs)[3] or may be an opsonin, a molecule such as complement or IgG that coat particles, targeting them for elimination by phagocytosis. Following target binding, receptors are recruited from outside the initial region of contact and, by sequential receptor-ligand interactions, guide actin-based pseudopods around the particle (Fig. 1). Although all phagocytosis requires actin polymerization, in some cases the phagocyte extends pseudopods to engulf the particle while in others the particle appears to sink into the cell. These morphologies are exemplified by IgG and complement opsonins, respectively.[4] The resulting phagosome has a tightly apposed membrane that conforms to the shape of the target and undergoes a series of fusion events that change its

Molecular Mechanisms of Phagocytosis, edited by Carlos Rosales. ©2005 Eurekah.com and Springer Science+Business Media.

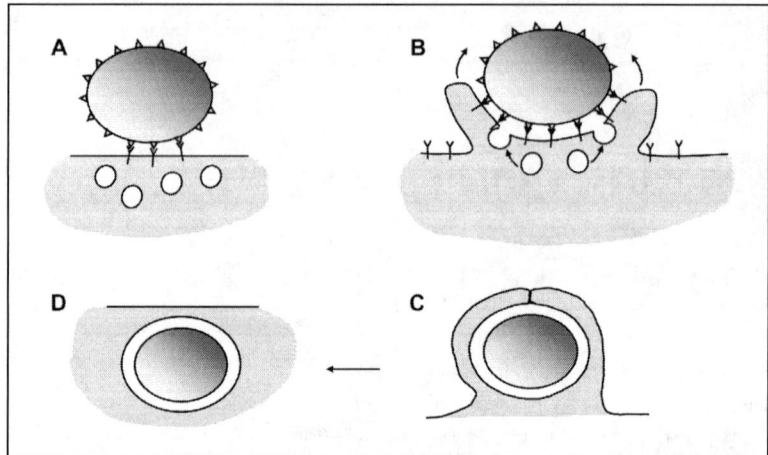

Figure 1. Overview of phagocytosis. A) Particles bind to cell surface receptors, cross-linking them and initiating signaling networks. Ruffling results in formation of endocytic vesicles. B) Sequential receptor-ligand interactions guide actin-based pseudopods around the particle. Insertion of VAMP3 positive vesicles into the forming phagosome provide membrane for pseudopod extension. Note that pseudopods extend beyond the membrane surface for IgG-mediated phagocytosis but complement-opsonized particles sink into the cell. C) Pseudopods fuse at the apex of the particle, enclosing the target in a phagosome (D).

composition. It transiently acquires markers of early (Rab 5, early endosome antigen 1, transferrin receptor) and late endosomes (Rab7 and mannose 6 phosphate receptor) before maturing into a phagosolysosome characterized by low pH (4.5-5.5) and the presence of lysosomal enzymes.[5] It should be noted that some bacteria subvert this generic sequence to survive and grow within the macrophage. Indeed, much of our information on phagososmal maturation has been derived from studies of macrophage pathogens.[6]

The type of receptor engaged will dictate subsequent signaling events. For example, IgG-mediated phagocytosis stimulates an inflammatory response, including activation of the respiratory burst for microbicidal killing and upregulation of inflammatory genes, including TNF-α, IL-1, and IL-12. In contrast, complement-mediated uptake is relatively silent, eliciting no burst nor activating pro-inflammatory genes. Some pathogenic bacteria, most notably *Mycobacterium tuberculosis*, exploit this pathway, entering macrophages via complement receptors and proliferating within this relatively safe environment. The ability of pathogens to disrupt the normal delivery of phagocytosed material to lysosomes is a large and diverse area of research and the reader is directed to several recent reviews.[6-9]

The IgG-dependent signal transduction cascade can be broadly divided into three stages, (1) target binding to the phagocyte cell surface with subsequent FcγR cross-linking, (2) proximal signaling events including tyrosine phosphorylation of the FcγR and (3) more distal events involving the activation of protein kinase C, mitogen-activated protein kinase (ERK), and transcription factors. Due to space considerations, this review focuses on the involvement of phospholipases in signaling events downstream of FcγR phosphorylation. However, excellent reviews on phagocytic signaling are available.[10-13]

Phospholipases

Phospholipases are enzymes that hydrolyze phospholipids, generating bioactive products that act as membrane detergents, enzyme cofactors, or second messengers. Phospholipids are built up on a glycerol or sphingosine backbone (Fig. 2). In higher eukaryotes, saturated and

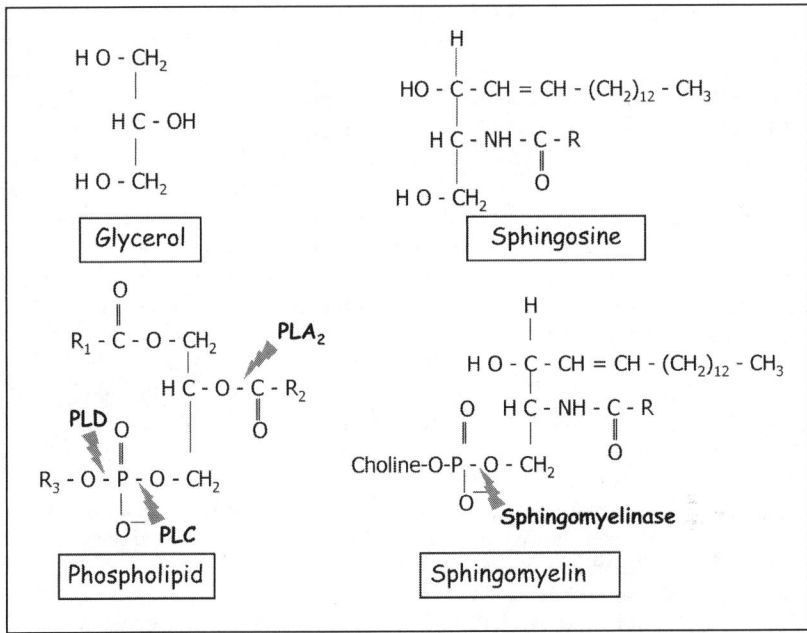

Figure 2. Phospholipase specificity. Phospholipases hydrolyze phospholipids that are built up on either a glycerol or sphingosine backbone. In mammalian membranes, the glycerol R_1 fatty acid is saturated and the R_2 is unsaturated. The R group in sphingosine is saturated. Sphingomyelinase and phospholipase C cleave at the same relative position on the sphingosine and phosphatidylinositol backbones, respectively. The R_3 headgroup is either choline, ethanolamine, inositol, or serine.

unsaturated fatty acids are esterified in the sn-1 and sn-2 positions of glycerol, respectively. The sn-3 position contains a phosphodiester linked headgroup: choline, serine, ethanolamine, or inositol. The position of cleavage on the glycerol backbone identifies the phospholipase family and generates unique products, many of which have second messenger function. Of particular interest in phagocyte function are (1) the arachidonic acid (AA) released by phospholipase A_2 (PLA_2), (2) inositol trisphosphate (IP_3) released by phosphatidylinositol-specific phospholipase C (PI-PLC), and (3) diacylglycerol (DAG) produced directly by PI-PLC or by the sequential action of phospholipase D (PLD) and phosphatidate phosphohydrolase (PAP) on phosphatidylcholine (PC). In contrast to these compounds which support phagocytosis, the ceramide product of sphingomyelinase may provide a brake, inhibiting PLD and terminating ingestion. Although not phospholipases, the phosphoinositide kinase family of enzymes phosphorylate the phosphatidylinositol ring at position 3, 4, or 5, generating species that function at several steps during phagocytosis. This review will address each class of phospholipase individually, starting with those more proximal to the receptors and ending with the sphingomyelinases.

Phospholipase C

Phosphatidylinositol-specific phospholipase Cs hydrolyze PI to the DAG and IP_3 second messengers which facilitate membrane targeting of signaling enzymes through lipid binding and increases in intracellular calcium (Ca_i), respectively.

PI-PLC are modular enzymes, with each of the 11 known isozymes sharing common pleckstrin homology (PH), flexible EF hand, and catalytic X and Y domains plus a C2, Ca binding, domain.[14] Additionally, different family members contain unique targeting domains, which place them in one of the subdivisions (β, δ, γ and the recently described ϵ).[14] Results from PLC null mice suggest that PLC β3 and PLCγ2 are both involved in immune signaling.[15] Neutrophils from PLCβ3 null mice exhibit defects in f met-leu-phe (FMLP)-stimulated respiratory burst and protein kinase activation.[15] PLCγ2 knock out animals have defects in numerous aspects of Fcϵ receptor signaling but notably no defect in IgG-mediated phagocytosis.[16]

In resting cells, PI-PLC is cytosolic, accumulating at the membrane upon receptor cross-linking.[14,17,18] Targeting may be via binding of the PH domain to the PI(4,5)P$_2$ (PIP$_2$) or PI(3,4,5)P$_3$ that is generated by the activation of PI kinases in response to cell stimulation.[14] Subsequent metabolism of these polyphosphoinositides results in release of the membrane-associated PLC and downregulation of the signaling network.

A two step translocation/activation model seems to account for the regulation of both PLC δ and PLC γ. For PLC-δ this has been dubbed the "tether and fix" model, whereby the enzyme is tethered to the membrane via binding of its PH domain to PIP$_2$ and fixed in place by binding of its C2 domain to the membrane. This fully activated form hydrolyzes PIP$_2$ with loss of the docking molecule, resulting in PLC dissociation from the membrane.[19] Alternatively, and particularly relevant to phagocyte signaling, is the localization and activation of PLCγ isozymes to the immunoreceptor tyrosine-based activating motif (ITAM) of immunoglobulin receptors.

Several lines of evidence support a model in which PLCγ is activated during IgG-mediated phagocytosis. These include the formation of IP$_3$ and DAG in IgG-stimulated neutrophils and monocytes,[20] tyrosine phosphorylation of PLC-γ1 and PLCγ2 upon FcγR cross-linking in human monocytes[21] and neutrophils,[22] respectively, the localization of PLCγ2 to phagosomes in RAW 264.7 macrophages,[18] and the loss of the IgG-induced Ca flux in PLC-γ2-null mice.[16] This latter report provides evidence that PLC-γ2 is the major PLC isoform linked to FcγR signaling. Interestingly, this study also demonstrated that *IgG-mediated phagocytosis* was unaffected by the loss of PLC-γ2, supporting earlier work with neutrophils and macrophages that assessed the effects of Ca depletion on phagocytosis. EGTA depletes Ca$_i$ to below detectable levels (< 2 nM).[23] Stimulation of Ca-depleted neutrophils or macrophages with IgG-opsonized targets elicited no detectable formation of IP$_3$, no rise in Ca$_i$, and no decrease in either the rate or extent of phagocytosis.[24,25] Taken together, these data suggest that the PI-PLCγ2 activated upon FcγR cross-linking signals for processes ancillary to phagocytosis, i.e., respiratory burst. How is PLCγ activated?

One of the earliest events following FcγR cross-linking is the src-dependent phosphorylation of the ITAM motifs in FcγRIIa tail or the γ chain of FcγRI and III (Fig. 3A).[11] ITAM phosphorylation generates phosphotyrosine docking sites for SH2 domain-containing proteins, including PLCγ, the tyrosine kinase Syk, and the regulatory subunit of PI3K.[26,27] In vitro studies using isolated ITAMs revealed that PLC-γ1 binds to phosphorylated ITAMs[26] and the p85 subunit of PI3K.[27] Tyrosine phosphorylation of PLC, either by src or syk, enhances its activity although maximal activation requires binding of PI(3,4,5)P$_3$ to either the PH or C-terminal SH2 domain[28] (Fig. 3A). The binding of PI3K to the phosphorylated ITAM can result in focal production of PI(3,4,5)P$_3$, providing a docking site for PLCγ. Thus, a picture emerges in which the ITAMs act as a scaffold to localize multiple enzymes for the targeting/activation of signaling enzymes, including PLC (Fig. 3). A fine level of control can be placed on the PLC by regulating the level of ITAM phosphorylation, the presence of PI3K in the complex, and/or the level of the PLC substrate, PIP$_2$, in the membrane.

Perhaps the best studied target of PI-PLC action is protein kinase C (PKC), a family of serine/threonine kinases that contain C1 and C2 domains that bind Ca and DAG, respectively. Ca/DAG binding targets the kinase to membrane complexes, allowing for signal propagation and subsequent respiratory burst.[29] The respiratory burst, which accompanies IgG-mediated phagocytosis, can also be activated by the DAG analog PMA. Moreover, Ca-depletion, PI-PLC inhibitors, and inhibitors of the Ca-dependent PKC (cPKC) isoforms block the burst in macrophages.[25,30] This argues for a role for both PLC and a cPKC isoform in activation of the

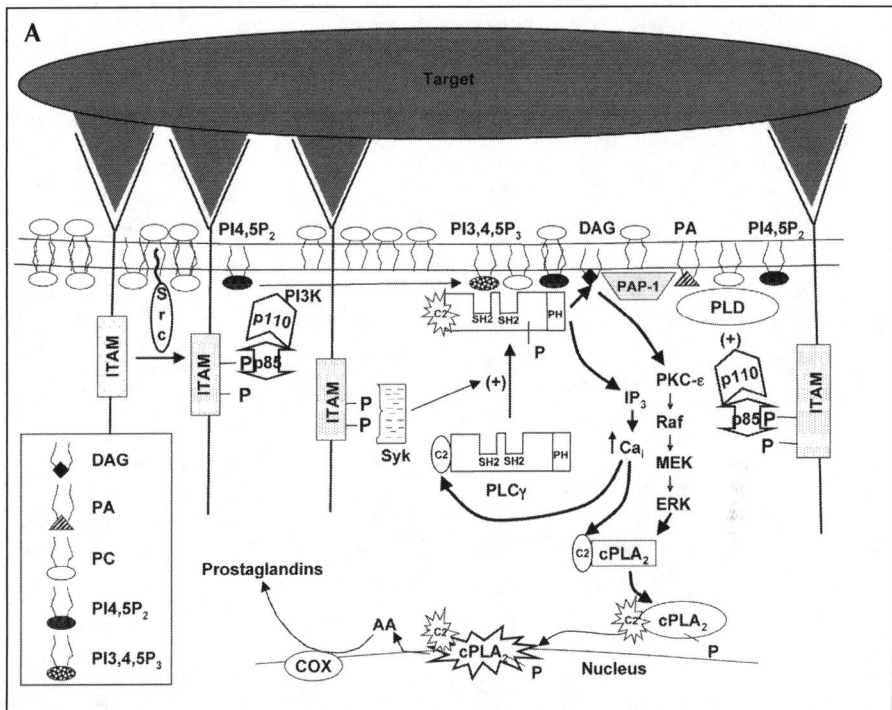

Figure 3. A) PLCγ contributes to activation of cPLA₂ during FcγR-mediated phagocytosis. FcγR cross-linking results in the phosphorylation of the ITAMs by a src family kinase member. The phosphorylated ITAMs form a docking site for several proteins, including Syk and p85/p110 PI3 Kinase. PI3K activation results in the formation of PI(3,4,5)P₃ which can bind to the C terminal SH2 domain of PLCγ, localizing it to the phagosome. Additionally, Syk phosphorylates PLCγ, enhancing its activity. Activated PLCγ hydrolyzes PI(4,5)P₂ to generate IP₃ and DAG. DAG, generated by PLC or the combined action of PLD and PAP-1, activates the novel PKC-ε, which is upstream of ERK. The combined effects of the rise in Ca$_i^{2+}$ and phosphorylation by ERK activate cPLA₂ to release AA for production of prostaglandins. Figure continued on next page.

respiratory burst. Studies in cell lines suggest that PKC-β mediates the burst in neutrophils while PKC-α serves this function in macrophages.[25,31] Additionally, PKC-α has been localized to IgG-containing phagosomes by immunofluorescence and biochemical analysis.[25,32] Not unexpectedly, PKC-α cannot translocate to phagosomes in Ca-depleted cells, arguing that both the IP₃ and DAG products of PI-PLC are necessary for cPKC translocation/activation under physiological conditions.[25]

The mechanism by which PKC modulates NADPH oxidase is likely to be the phosphorylation of the p47^{phox} subunit of the oxidase, with one study reporting a direct interaction between p47^{phox} and cPKCs βI and βII.[33] Based on the literature, the following model can be constructed. The binding of an IgG-opsonized particle activates PI-PLCγ2 to release of IP₃ and DAG in the region of the particle (Fig. 3B). Released Ca binds to the C2 domain of cPKC α or β, directing it to the membrane where it binds DAG via its C1 domain[29] (Fig. 3B). The ensuing conformational change activates the PKC, facilitating its phosphorylation of p47^{phox}. Phosphorylation causes a conformational change in the p47^{phox} that permits assembly of the active oxidase complex.[34] Retention of the active PKC at the membrane may result in hyperphosphorylation of

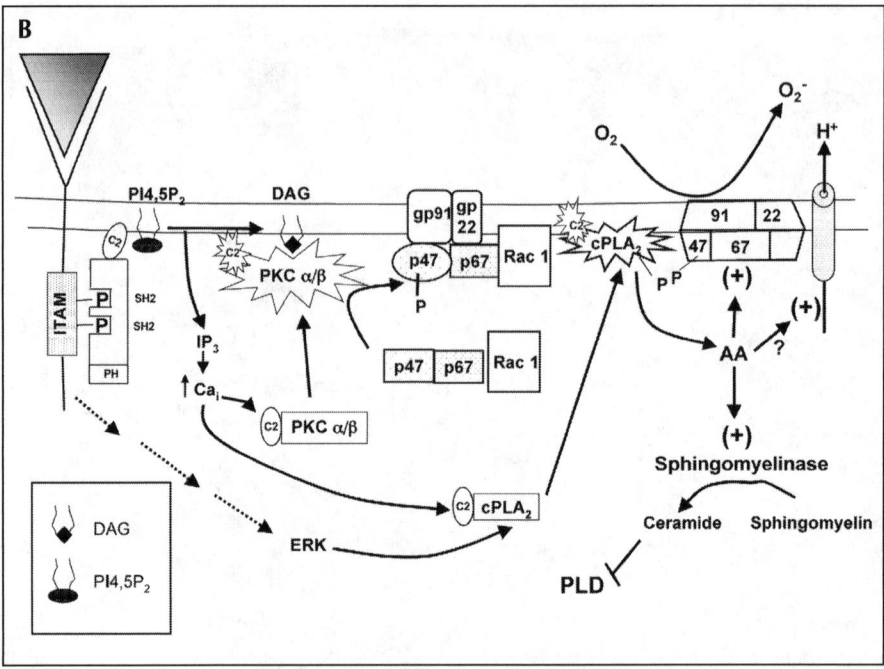

Figure 3, continued. B) PLCγ contributes to activation of NADPH oxidase during FcγR-mediated phago-cytosis. FcγR cross-linking results in the phosphorylation of the ITAMs by a src family kinase member. The phosphorylated ITAMs form a docking site for several proteins, including PLCγ. PKC α/β, activated by the rise in [Ca$_i$] and release of DAG, phosphorylates p47^{phox}, resulting in association of the p47/p67/Rac1 complex with the gp91/gp22 membrane components of the NADPH oxidase. AA, released by cPLA$_2$, then activates NADPH for production of superoxide (O$_2^-$). There is also evidence that AA regulates a proton channel that increases the efficiency of the oxidase. AA, released by cPLA$_2$ may also activate sphingomyelinase to produce ceramide which inhibits PLD. Inhibition of PLD may block PKC-ε activation, thus depressing ERK activation, cPLA$_2$ phosphorylation, and the respiratory burst.

p47^{phox} resulting in retranslocation of the p47/p67^{phox} complex back to the cytosol, effectively disassembling the oxidase.[35] Although consistent with the literature, it should be noted that the PKC isoforms involved have yet to be definitively identified and that the mechanism of regula-tion may be different for neutrophils and monocyte/macrophages.

In summary, activation of PI-PLC accompanies IgG-mediated phagocytosis and its prod-ucts aid in the targeting of cPLA$_2$ (see below) and cPKC isoforms to the phagosome for assem-bly of the NADPH oxidase and oxidative killing. However, the ability of DAG to enhance phagocytosis and the inhibition of uptake seen with PKC inhibitors or dominant negative PKC fragments suggests that a DAG-sensitive PKC is necessary for phagocytosis.[25,36,37] If ph-agocytosis is Ca-independent, what is the source of this DAG? Emerging evidence suggests that DAG release may be dependent on activation of phospholipase D.

Phospholipase D

Phospholipase D (PLD) hydrolyzes PC to phosphatidic acid (PA) and choline.[38] The PA is then dephosphorylated to DAG by phosphatidate phosphohydrolase (PAP-1)[39] (Fig. 3A). In the presence of primary alcohols, PLD catalyzes a transphosphatidylation reaction, resulting in the production of nonsignaling phosphatidylalcohols (PEt or PBu, phosphatidylethanol or phosphatidylbutanol). As PA is itself a second messenger (see below), primary alcohols not only prevent PA-mediated signaling but also block the production of PLD-derived DAG. To date, two mammalian PLDs have been described, human PLD1 and mouse PLD2. Human PLD1 has two splice variants, the 124 kDa PLD-1a and PLD-1b, a shorter variant lacking 38 amino acids in a nonregulatory region. Both PLD1 and PLD2 have similar regulatory regions, including PX and PH domains and an additional conserved region that binds PIP_2.[38] PLD1a/b are regulated by cPKC (α/β), ARF6 (a small molecular weight G protein involved in membrane trafficking), and members of the Rho family of small molecular weight G proteins. PLD2 has a high constitutive activity.[38,40] However, coexpression of PLD2 with either PKC-α or ARF6 in human embryonic kidney (HEK) cells increases its activity, suggesting at least some level of regulation by ARF6 and PKC-α.[38] Through their tight association with PIP_2, both PLD1 and PLD2 may localize to regions of increased PIP_2 expression, with PLD1 activity being more finely controlled by interactions with additional effectors.

Phagocytosis of complement and IgG-opsonized particles in neutrophils is accompanied by production of IP_3 and DAG, implicating activation of PI-PLC.[41] However, as phagocytosis is not altered upon Ca-depletion (although IP_3 release is blocked) PI-PLC may not be involved.[41] The fact that DAG is released from Ca-depleted cells at a slower rate than PA is consistent with a PA→DAG pathway whereby PA is released by PLD and subsequently metabolized to DAG.[42,43] The production of PEt verified the activation of PLD in complement-mediated phagocytosis.[41]

Working with human monocytes, Kusner et al demonstrated that PLD is activated and necessary for phagocytosis of serum-opsonized zymosan (engaging the complement receptors), IgG-opsonized beads (FcγR-mediated uptake), or *Mycobacterium tuberculosis*.[44,45] These biochemical results are supported by molecular studies demonstrating that GFP-conjugated PLD2 concentrates at IgG-containing phagosome in macrophages (Ueyama et al submitted, Cheeseman and Lennartz, unpublished observation). Localization may be the result of PIP_2 accumulation in the forming phagosome[18] and PLD may be activated by polymerized actin, similar to results obtained in vitro.[17] Interestingly, PA has been shown to stimulate actin assembly on isolated latex bead phagosomes,[46] suggesting a feedback loop in which localized PIP_2 production directs PLD docking which releases PA for actin polymerization. The filamentous actin thus formed feeds back to enhance PLD activity. How PA contributes to actin assembly is currently unknown.

In addition to its effects on actin, PA may modulate membrane trafficking. We propose a model in which PLD-mediated membrane ruffling facilitates insertion of vesicles into the forming phagosome for pseudopod extension (Fig. 4). Phagocytosis requires PLD and ARF6.[44,45,47] In mast cells, antigen stimulates membrane ruffling and exocytosis of secretory granules. PLD2 and ARF6 localize to these ruffles and 1-butanol (which prevents PA formation) blocks both ARF6 accumulation and membrane ruffling;[48] expression of dominant negative PLD2 blocks secretion.[19] Taken together, these data suggest that PLD is upstream of ARF6 and that PA is necessary for ruffling and subsequent secretion. Whether PA conversion to DAG is necessary for ruffling is not known. With respect to phagocytosis, membrane ruffling is apparent at sites of target uptake.[50-52] This ruffling may result in the formation of vesicles for membrane trafficking and/or in insertion of intracellular vesicles for pseudopod extension. We have shown that plasma membrane-derived vesicles accumulate beneath bound targets in an PLA_2 dependent manner (see below).[51] Additionally, Aues et al reported that wortmannin (a PI3 kinase inhibitor) blocks PLD and endocytosis in U937 macrophages,[46] implicating PI3K in the activation of PLD and endocytosis. The mechanism of this activation remains to be elucidated. That wortmannin inhibits IgG-mediated phagocytosis of large, but not small particles, suggests a block in pseudopod extension that may result from a defect in endocytosis.[53] Finally, in mast cells PLD1 is involved in translocation of vesicles while PLD2 is necessary for exocytosis.[49] Applying a similar rationale to phagocytosis, PLD1 may be involved in formation, and PLD2 necessary for the redistribution and/or exocytosis, of vesicles. Taken together, these data

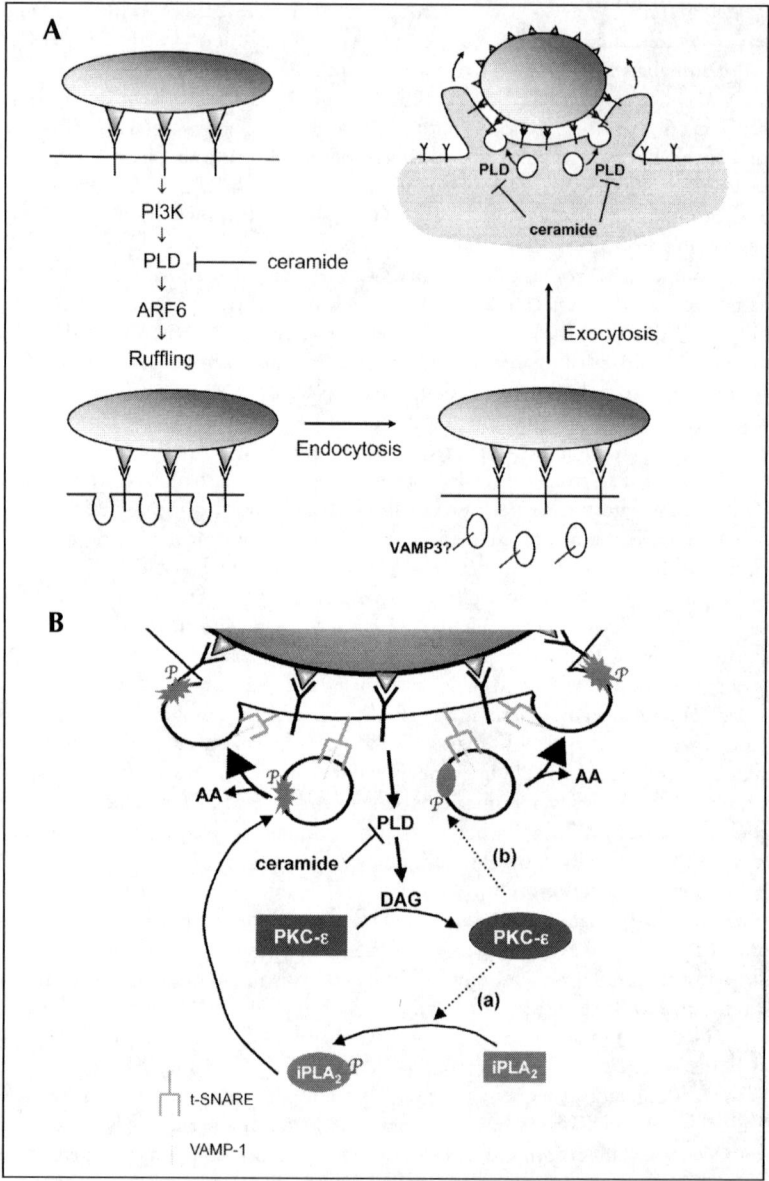

Figure 4. Model for PLD involvement in membrane redistribution during phagocytosis. A) Target binding recruits PI3K (see Fig. 3) which enhances PLD activity. PLD is upstream of the ARF 6 that is necessary for membrane ruffling and endocytosis. Endocytic vesicles expressing VAMP3 then fuse into the forming phagosome in a process requiring iPLA$_2$. Both formation and exocytosis of vesicles requires PLD activity. Ceramide accumulation may inhibit PLD, thus turning off the phagocytic response. B) Possible mechanisms of iPLA$_2$ modulation by PKC-ε. a) PKC-ε, activated by PLD, may phosphorylate cytosolic iPLA$_2$, facilitating its association with VAMP3 positive vesicles. Release of AA by activated iPLA$_2$ results in exocytosis and extension of pseudopods. b) Alternatively, iPLA$_2$ may be constitutively associated with the vesicles. PKC-ε may phosphorylate and activate the membrane-bound iPLA$_2$ for production of AA and exocytosis. Inhibition of PLD by ceramide may block exocytosis and elongation of pseudopods.

are consistent with the following pathway: PI3K→ PLD1→ ARF6 → ruffling→ endocytosis → redistribution of vesicles → PLD2 → exocytosis → pseudopod extension (Fig. 4A). The vesicles are likely to be the VAMP3 positive recycling endosomes that insert into IgG-containing phagosomes.[54] However, whether they contain ARF6, PLD1, and/or PLD2 remains to be determined. Additionally, efforts to understand how PLD facilitates membrane fusion (either for endocytosis or exocytosis) should produce exciting results.

Lipid Phosphate Phosphohydrolase (LPP)

As their name implies, LPP remove phosphates from a variety of lipid substrates, including PA, sphingosine 1-phosphate, and ceramide 1-phosphate.[39] Of particular interest to phagocytosis are the LPP subfamily of phosphatidate phosphohydrolases (PAP), enzymes that hydrolyze PA to DAG. The specific isoforms involved in immune signaling have not been identified although work by Sciorra and Morris have shown that PLD2 and PAP2b colocalize to detergent resistant membranes (i.e., lipid rafts or caveolae) and act in concert to produce DAG in these membrane domains.[43] Although this work was done in human embryonic kidney cells (HEK) and Swiss 3T3 mouse fibroblasts, these results may be applicable to macrophages and phagocytosis for the following reason: PLD2 is the major PLD isoform expressed in HEK and 3T3 cells and it is found in membrane rafts where PAP2b is also localized. Similary, PLD2 is the only isoform detected by Northern blots of RAW 264.7 mouse macrophages (Cheeseman and Lennartz, unpublished). Additionally, FcγR are found in detergent resistant membranes upon ligation where they are phosphorylated by Lyn, a Src family member.[55] Thus, PLD2, PAP2b, and FcγR with its associated proteins may concentrate in lipid rafts upon target binding, thus increasing the efficiency of phagocytic signal transduction. How might this occur?

The sequential action of PLD and PAP2b on PC generates DAG.[43] As FcγR-mediated phagocytosis is Ca-independent but requires PKC, activation of PLD may provide the DAG necessary for localization of PKC to phagosomes. What PKC isoform might this be? We recently reported that the PKC-ε is required for efficient IgG-dependent phagocytosis in mouse macrophages.[37] Similarly, PKC-ε null mice are more sensitive to bacterial challenge than their wild type counterparts, suggesting that PKC-ε may also be necessary for phagocytosis of pathogens.[56]

Membrane localization of PKC-ε is regulated by DAG but does not require Ca. This conclusion is based on reports that DAG/PMA stimulates accumulation of PKC-ε at the plasma membrane in COS-7 cells[57] and RAW 264.7 mouse macrophages (Lennartz, unpublished) and that EI-150, a DAG antagonist, blocks localization of GFP-PKC-ε to membranes[57] and forming phagosomes (Ueyama, unpublished observation). Additionally, Ca depletion does not alter the accumulation of PKC-ε at phagosomes.[25] Finally, PLD2 concentrates at the phagosome in macrophages with a timecourse similar to that of PKC-ε (Ueyama et al submitted; Cheeseman and Lennartz, unpublished observation). It is of interest to note that PKC-ε is the only isoform that localizes to phagosomes in Ca-depleted cells.[25] It is also the only isoform with an actin binding domain.[58] Binding of PKC-ε to DAG exposes this actin binding domain and enhances the catalytic activity of the enzyme is 4-fold.[59] We have shown that the isolated regulatory domain of PKC-ε (εRD) localizes to phagosomes and accumulates actin although phagocytosis is depressed.[37] Thus, PLD-derived DAG may recruit PKC-ε to nascent phagosomes but the catalytic activity of PKC-ε is necessary to propagate the ingestion signal. Identification of PKC-ε target proteins should provide insight into how it contributes to phagocytosis.

Phospholipase A₂ (PLA₂)

We next turn to the PLA₂ family members. They are positioned at an intermediate level in the phagocytic signaling network, being activated downstream of PLC and PLD and contributing to several different outputs, including ingestion, membrane redistribution, and NADPH oxidase activation.

The phospholipase A_2 (PLA$_2$) superfamily consists of 12 family members comprising secretory (sPLA$_2$) and cytosolic family members.[60,61] The intracellular forms are either Ca-dependent (cPLA$_2$, Class IV) or Ca-independent (iPLA$_2$, Class VI). PLA$_2$ hydrolyze PC or phosphatidylethanolamine (PE) to release a free fatty acid and lysophospholipid. Arachidonic acid (AA) is present at unusually high levels in the phospholipids of inflammatory cells. It is released upon PLA$_2$ activation to participate in signaling networks as a second messenger or to be metabolized into the eicosanoid inflammatory mediators that modulate the immune response. IgG-mediated phagocytosis results in the preferential release of AA over other fatty acids[23] via activation of cPLA$_2$ and iPLA$_2$.

iPLA$_2$

In macrophages, FcγR-mediated phagocytosis is Ca-independent and requires AA that is released from PE.[23] iPLA$_2$ inhibitors block AA release and phagocytosis which can be restored upon addition of exogenous AA, suggesting a critical role for AA in ingestion. Preventing AA metabolism does not alter phagocytosis arguing that AA per se, rather than one of its metabolites, mediates uptake.[62] Ultrastructural examination of cells in which iPLA$_2$ was inhibited revealed an accumulation of vesicles beneath bound targets. No such structures were seen when AA was added to the inhibited cells, suggesting that AA may be facilitating the insertion of these vesicles into the forming phagosome.[51] There is considerable precedence for a role for PLA$_2$ in membrane fusion that would be consistent with this model, particularly those supporting the involvement of PLA$_2$ in fusion between intracellular organelles or during exocytosis.[63,64] Recent molecular studies demonstrated that VAMP3-containing vesicles, presumably recycling endosomes, are indeed inserted into phagosomes.[54] Taken together, this work suggests that iPLA$_2$ associates with endosomes and that its AA product facilitates membrane trafficking during phagocytosis (Fig. 4B). The identification of iPLA$_2$ activity on isolated endosomes provides support for this model.[65] Additionally, reports that exogenous PLA$_2$ stimulates degranulation in mast cells[66] and exocytosis in parotid acinar cells,[67] that PLA$_2$ inhibitors block endosome fusion,[68,69] and that AA can overcome this block[64] strengthen the model. Still under investigation, however, are the mechanisms by which the PLA$_2$ is activated.

In human monocytes, membrane-associated iPLA$_2$ activity increases on membranes during phagocytosis and in response to the PKC activator, PMA.[70] If this class of PLA$_2$s is not regulated by phosphorylation or binding to second messengers, then translocation to the membrane substrate may constitute "activation".[60,61]

However, the fact that iPLA$_2$ lies downstream of PKC in the phagocytic signaling pathway raises the intriguing possibility that PKC may phosphorylate iPLA$_2$ and contribute to its association with endosomes.[70] Whether this is via direct activation of the enzyme at the membrane or translocation of a cytosolic pool to the forming phagosome has yet to be determined (Fig. 4B). It should be noted that direct phosphorylation of iPLA$_2$ has not been reported although the sequence of an iPLA$_2$ from rabbit kidney indicates several putative PKC phosphorylation motifs.[71] The modulation of iPLA$_2$ by PKC has also been reported in zymosan-stimulated macrophages.[72] Similar to IgG, zymosan increases the activity of iPLA$_2$ in the membrane fraction of P388D1 macrophages.[72] Also, AA release is increased or decreased by activators or inhibitors of PKC, respectively,[72] as determined by quantifying the production of the AA metabolite prostaglandin D2 in response to zymosan.[72] Taken together, these data suggest that macrophages have a PKC→ iPLA$_2$ pathway that can be engaged by multiple stimuli, leading to phagocytosis and/or AA metabolism. iPLA$_2$ may be activated by PKC-dependent phosphorylation in a manner analogous to the well-documented modulation of cPLA$_2$ by ERK-dependent phosphorylation (see below). However, it remains to be seen if zymosan phago-cytosis requires AA release and whether IgG-mediated prostaglandin production is iPLA$_2$ dependent. Unfortunately, with the exception of bromoenol lactone, there are no reagents (i.e., clones, antibodies) available for this class of PLA$_2$s. Detailed studies on the activation/regulation/ localization await the development of specific reagents.

cPLA$_2$

There are three cPLA$_2$ isoforms: α, β, and γ.[73] cPLA$_2$-α is the isoform expressed in hematopoietic cells and thus is the subject of this discussion (hereafter referred to as cPLA$_2$). As phospholipids are the cPLA$_2$ substrate, translocation to membranes is critical for their signaling. cPLA$_2$ contains a C2 domain and is translocated in response to Ca-ionophore and/or stimuli that raise Ca$_i$.[73] This rise in Ca is necessary, but not sufficient, for translocation to particulate stimuli; phosphorylation on Ser-505 by ERK1/2 provides the full activation signal.[73-76] Although activated during phagocytosis, as evidenced by translocation to membranes,[77] there is no evidence that cPLA$_2$ is necessary for particle uptake. Indeed, the Ca-independence of IgG-dependent phagocytosis argues against the involvement of cPLA$_2$, although its activity in Ca-depleted cells has not been studied. However, phagocytosis initiates a network of events that results not only in the uptake of the particle, but also in stimulation of the respiratory burst and release of eicosanoids, which do seem to require cPLA$_2$.

cPLA$_2$ and Respiratory Burst

Early pharmacological studies demonstrated that PLA$_2$ inhibitors decreased IgG dependent respiratory burst and that AA could bypass the inhibition.[78,79] As PLA$_2$ hydrolysis of membrane phospholipids generates products that change the physical characteristics of the membranes, it was not until Dana et al demonstrated that AA stimulated NADPH oxidase in a cell free system that AA was acknowledged as a direct modulator of NADPH oxidase.[80] In these studies it was shown AA was not involved in oxidase assembly but was necessary for activation of the assembled complex (Fig. 3B). More recent work supports a model in which AA is necessary for the respiratory burst stimulated by opsonized zymosan (OZ), PMA, and the chemotactic factor, FMLP.[81] These studies were carried out in the myeoloid cell line PBL-D, engineered to lack cPLA$_2$. PBL-D cells are incapable of mounting a burst in response to OZ, PMA, or FMLP. The fact that the block can be overcome by exogenous AA[81] and that cPLA$_2$ is detected in pull-down assays using GST-p47^{phox} or GST-p67^{phox} suggests that cPLA$_2$ directly binds to one or more components of the oxidase complex, generating a localized release of AA for oxidase activation.[82] Although this model fits the data generated in the PBL-D cell line, it should be noted that peritoneal macrophages from cPLA$_2$-deficient animals have a normal burst in response to PMA and OZ although AA release and metabolism is depressed.[75] Whether this discrepancy is due to compensation in the null animals, a fundamental difference between the myeloid precursors and differentiated macrophages, or a species difference (PBL are human-derived cells), it reminds us that care must be taken in generalizing results across systems.

The exact mechanism by which AA regulates the oxidase is not known. However, the observations that AA stimulates a Zn-inhibitable proton channel and that Zn blocks respiratory burst are consistent with a model in which cPLA$_2$ releases AA which activates proton channels.[83,84] The net result of membrane proton channel activity would be a transient alkalinization of the cytosol, leading to more efficient electron transfer and respiratory burst. AA regulation of other cation channels has been reported, providing precedence for such a model.[85,86] As regulation of proton channels does not explain the activity of AA in cell free systems, this fatty acid may have several effects in intact cells, actions that may work in concert or be selective for different stimuli or cell types.

cPLA$_2$ and Eicosanoid Production

A role for cPLA$_2$ in eicosanoid production was suggested by studies demonstrating trans-location of GFP-cPLA$_2$ to the nucleus and AA release in response to A23187 in Sf9 insect cells.[87] Although the Sf9 system is artificial, the results complement studies in neutrophils, which are refractory to molecular manipulation.[77] In a scenario parallel to that postulated for regulation of NADPH oxidase, nuclear localization of cPLA$_2$ would release AA at the site of cyclooxygenase and lipoxygenase action, providing the substrate for the production of prostag-landins and leukotrienes, respectively.[76] Activation of ERK and the rise in Ca$_i$ would provide "fine control" of the cPLA$_2$ activity, regulating catalytic activity by phosphorylation of Ser505 and translocation by the extent of the rise in Ca$_i$.[61,76]

As IgG-dependent phagocytosis elicits a rise in Ca$_i$, ERK activation, release of AA, and eicosanoid production,[23,88,89] it is reasonable to postulate that a similar mechanism operates during IgG-mediated phagocytosis (Fig. 3A). However, nuclear translocation of cPLA$_2$ in re-sponse to IgG-opsonized targets has yet to be shown. Additionally, the question of how cPLA$_2$ is targeted to both the plasma membrane (for oxidase activation) and the nuclear membrane (for eicosanoid production) remains to be answered. Is there differential phosphorylation of secondary sites, perhaps by different PKC isoforms? There are several PKC consensus sites in cPLA$_2$.[76] What are the PKC isoforms? What are the phosphorylation sites? How is this con-trolled? Alternatively, does some fraction of the cPLA$_2$ associate with an ERK scaffold for transit to the nucleus? Indeed, ERKs form scaffolds with other enzymes of the ERK activation pathway[90] and ERKs are targeted to the nucleus where they modulate gene expression.[91] Or does the extent of the Ca$_i$ rise or level of ERK activation differentially regulate cPLA$_2$ localiza-tion? The answers to these questions will advance our understanding of the role of phospholi-pase A$_2$ in phagocyte signaling.

Lysophospholipids

Because the majority of PLA$_2$ studies have focused on the AA product of PLA$_2$, the in-volvement of the lyso-PC product in immune signaling has received little attention. However, several reports suggest that these lipids may affect phagocyte function. When incubated with rabbit alveolar macrophages, lyso-PC enhances phagocytosis of IgG-opsonized targets.[92] Con-versely, PC itself depresses that function. Similar findings were reported with lysophosphatidic acid (lyso-PA), the product of the sequential action of PLD and PLA$_2$. Microinjection of lyso-PA into mouse bone marrow-derived macrophages enhances phagocytosis of IgG-opsonized latex beads.[93] The lyso-PA mediated enhancement was similar to that obtained by microinjecting consitutively active Rac1, a small molecular weight G protein required for IgG-mediated phagocytosis and known to be activated by lyso-PA.[94-96] Although other macrophage func-tions were not assessed, this result raises the intriguing possibility that activation of PLA$_2$ may provide a generalized signal to the macrophages to upregulate their inflammatory response. If lyso-PC/PA activation of phagocytes is shown to be a general phenomenon, it would provide a very efficient mechanism by which both PLA$_2$ products facilitate inflammation, one (AA) by generating soluble secondary messengers/inflammatory mediators and the other (lyso-PC) by mediating the membrane/cytoskeletal changes necessary for phagocytosis and chemotaxis.

Sphingomyelinases

Once ingestion, respiratory burst, eicosanoid release, and gene regulation have been turned on in response to a pathogen how does the system return to baseline? This aspect of phagocy-tosis has received much less attention than activation of the network. However, interesting findings implicate the sphingomyelinases as major players in generating the "off" signal.

Sphingomyelinases hydolyze sphingomyelin in a PLC-like reaction that results in the pro-duction of ceramide and phosphorylcholine (Fig. 2). At least 5 classes have been defined but few are cloned and research in the field relies heavily on pharmacological and biochemical techniques.[97] The ceramide product has saturated side chains that allow it to pack tightly into membrane rafts, detergent insoluble membrane domains that organize signaling complexes. Additionally, ceramide modulates the activity of a variety of signaling molecules, including the PKC-ε and PLD involved in phagocytosis. Thus, ceramide is a dual action molecule that can function either as a second messenger or to alter the structural architecture of the cell membrane.

In vitro assays demonstrate that ceramide modulates the activity of a number of signaling molecules, activating some (ceramide-activated protein phosphatase and PKC-ζ) and inhibiting others (PKC-ε and PLD).[98-101] The intrinsic hydrophobicity of ceramide dictates that it remains in the membrane in which it was formed, necessitating movement of cermide targets to the memebrane.

With respect to phagocytosis, PKC-ε and PLD are necessary for uptake of IgG-opsonized targets,[37,102] PKC-ζ contributes to the unique "delayed phagocytosis" characteristic of *Helicobacter pylori*.[103] Measurements of ceramide release reveal that it is produced relatively late during phagocytosis.[101,104] Thus, the gradual accumulation of ceramide may provide the "off" signal for IgG-mediated phagocytosis (by inhibiting PKC-ε and/or PLD) and an "on" signal for *H. pylori* ingestion (delayed activation of PKC-ζ). The fact that PKC-ζ accumulates on *H. pylori* phagosomes supports a model in which local production of ceramide is the target for PKC-ζ localization.[103]

Ceramide blocks several aspects of IgG-dependent signaling, including phagocytosis, PEt production (PLD), H_2O_2 release (NADPH oxidase), and activation of ERK1/2.[101,104-106] Occam's razor would predict that inhibition of a single receptor-proximal reaction may explain the plethora of ceramide effects. A model in which ceramide blocks PLD activity would be consistent with the available data. Inhibition of PLD blocks phagocytosis.[44,45,52,107] Our studies suggest that this inhibition is due to low PKC-ε activity.

With respect to NADPH oxidase, inhibition of respiratory burst may occur through ERK1/2. ERK phosphorylates cPLA$_2$ which is necessary for activation of NADPH oxidase.[81,108] ERK is regulated by a Ca-independent PKC that is activated upon FcγR ligation.[89] PKC-ε is the only novel isoform that translocates during phagocytosis[25] and thus is likely to be upstream of ERK (Fig. 3A). PKC-ε modulates ERK activity by interacting with Raf-1, a MAP kinase kinase.[100,109] In our model, ceramide inhibition of PLD would prevent PKC-ε localization and activation of ERK, thus depressing NADPH oxidase activity and H_2O_2 production (Fg. 3). In HEK cells, ceramide decreases insulin-like growth factor I (IGF-1) stimulated ERK activity.[100] The authors demonstrate that IGF-1 treatment depresses PKC-ε activity, PKC-ε/Raf-1 association, and PKC-ε/ERK interactions. However, these studies were done in intact cells, using immunoprecipiation assays to define the interactions. Thus, although the results are consistent with the conclusion that ceramide is directly blocking PKC-ε, the involvement of PLD in this system was not addressed and could easily account for the results. Banno et al examined the role of PLD in IGF-1 signaling in CHO cells. They reported that IGF-1 activates PLD and that ERK activation is suppressed by 1-butanol or expression of catalytically inactive PLD 1 or 2.[110] This work provides additional evidence that ceramide, acting through PLD, depresses ERK activity. Likewise, the ceramide effect on ERK in HEK cells could be mediated via inhibition of PLD. If a similar pathway is operational in macrophages, ceramide may exert all the observed effects on phagocytosis and respiratory burst by inactivating PLD. Although considerable evidence exists that ceramide inhibits PLD, the mechanism is unknown.

Although ceramide plays a central role in terminating the phagocytic response, the identity of the sphingomyelinase involved and how it is activated remain to be determined. One clue is provided by a study in L929 cells and their C12 derivative, which lacks cPLA$_2$. In parental cells, TNF-α stimulates AA release, which is followed by delayed production of ceramide.[111] In contrast, C12 cells do not generate ceramide in response to TNF-α ; reexpression of $cPLA_2$ partially restores the ceramide response.[111] Thus, one could speculate that $cPLA_2$, which is activated during IgG-mediated phagocytosis (see above), upregulates sphingomyelinase. A Mg^{2+} dependent neutral sphingomyelinase is one possible candidate as it is activated upon IgG stimulation of neutrophils and is one of the three sphingomyelinases responsive to TNF-α in L929 cells.[111,112] This model predicts that sphingomyelinase is one component of a secondary signaling pathway. AA, released upon $cPLA_2$/$iPLA_2$ stimulation, would bind to shuttle vesicles, the NADPH oxidase, and possibly other signaling enzymes. Any AA that is not consumed in these primary networks would be available to interact with the sphingomyelinase.

Sustained sphingomyelinase action would change the biophysical characteristics of the membrane as the ceramide coalesces into rafts. Upon cross-linking, $Fc\gamma R$ associates with detergent-insoluble membranes, where it localizes with lyn to initiate the signaling network.[113] At this time, rafts would serve as signaling platforms and support both phagocytosis and its associated responses (gene regulation, respiratory burst). With time, sphingomyelinase will be recruited/activated and the ceramide produced will concentrate in the rafts. At some point, the ceramide concentration will be high enough to inactivate PLD. One can imagine that this inactivation will be gradual and will lead to turning down, then off, the $Fc\gamma R$ signals. A report that a neutral sphingomyelinase is sequestered in membrane rafts containing the src kinase lyn in U937 cells provides evidence in support of this model.[114] To complete the cycle, the entire signaling complex would be internalized in the phagosome, with the $Fc\gamma R$ being recycled back to the membrane or being degraded.

In conclusion, a considerable amount of evidence exists for a role for sphingomyelinases/ceramide in termination of $Fc\gamma R$ responses. However, much of this evidence is circumstantial and the test of the model will be to determine if modulation of ceramide indeed downregulates PKC-ε, ERK, and/or $cPLA_2$ activity during phagocytosis. Also, the question remains: how does ceramide block PLD?

Lipid Kinases

No discussion of lipids and phagocytosis would be complete without considering the lipid kinases and their phosphoinositide products. Advances in the technology for studying these polyphosphoinositides, including the identification of lipid-binding domains (e.g., PKC-δ PH, Akt-PH, FYVE) that specifically bind different lipid kinase products (e.g., $PI(4,5)P_2$, $PI(3,4,5)P_3$, $PI(3)P$) and high resolution fluorescence imaging, have propelled this field forward in recent years.

Although an detailed discussion of these results is beyond the scope of this review, a few points bear mentioning. Wortmanin, a PI 3 kinase inhibitor, selectively blocks phagocytosis of large particles, suggesting that PI3K products are necessary for pseudopod extension.[53] This is consistent with reports demonstrating that phosphoinositides regulate virtually every aspect of membrane trafficking, including endocytosis, endosomal maturation, and phagocytosis, with different phosphoinositides involved in each process.[115] For phagocytosis, PIP_2 serves as a substrate for PI-PLC, an activator of PLD,[43] and to regulate actin polymerization. PIP_2 may subserve this latter function by binding to gelsolin, thus uncapping the barbed end of actin and facilitating actin polymerization at the phagosome.[116,117] $PI(3,4,5)P_3$ accumulates exclusively at the phagosome, suggesting a specific role in internalization or signaling, possibly the localization of PLCγ.[28] Additionally, a membrane permeant form of $PI(3,4,5)P_3$ induces actin accumulation and migration in neutrophils, consistent with a role in membrane movement, possibly pseudopod extention.[5] PI3P accumulates on phagosomes upon closure and is then retained for several minutes. Inhibition of PI3 kinase suppressed appearance of EEA1 (early endosome antigen) on the phagosome, suggesting that PI3P is involved in phagosome maturation.[5] These studies demonstrate the intricate involvement of phosphoinositides in phagosome formation and maturation, but merely scratch the surface of the literature. The reader is directed to recent reviews for an excellent overview of this exciting field.[5,115,116,118,119]

Conclusion

As our knowledge of phagocytic signaling grows, it is clear that the process is more a signaling network than a linear pathway. Phospholipase products contribute not only to the internalization of particles, but also to respiratory burst and gene regulation. The process is initiated by cross-linking of receptors, amplified by the generation of second messengers that impact on one or more responses, and then terminated. Lipid second messengers are now being appreciated for their complex involvement in all three of these phases. Arachidonic acid is one obvious example. Experiments detailed here demonstrate that this lipid acts early in the phagocytic process, contributing to pseudopod extension, later as an activator of NADPH oxidase and as the precursor to the eicosanoid inflammatory mediators, and finally as a potential effector of sphingomyelinase to shut off the phagocytic response. Likewise, the PA product of PLD can contribute to NADPH oxidase activation or vesicular trafficking and its DAG metabolite contributes to activation of several PKC isoforms for phagocytosis and gene regulation.[120] Finally, ceramide, produced relatively late may turn out to be the master regulator to turn off signaling at a very proximal point. However, this remains to be seen. The availability of molecular tools, high resolution imaging, and proteomics, and ongoing technical advances, is allowing us to shine a light into many of the current "black boxes" in our knowledge. The results, like those summarized here, will expand our horizons into unknown and exciting areas.

Note: I have tried to assemble the literature into testable models for the involvement of phospholipases in phagocytosis. Some models have more supporting data than others. I have tried to be as comprehensive as space allows. Any oversight of relevant literature is unintentional.

Definitions

Arachidonic Acid—20 carbon unsaturated fatty acid containing 4 double bonds. Product of phospholipase A_2 action on phosphatidylcholine or phosphatidyethanolamine which acts as a second messenger.

C2 domain—Calcium (Ca) binding domain of ~130 amino acids that facilitates the translocation and binding of proteins to membranes in response to increases in intracellular Ca.

Diacylglycerol (DAG)—Lipid product of phospholipase C; activator of protein kinase C (PKC).

Immunoreceptor tyrosine-based activating motif (ITAM)—Homologous sequence of amino acids in the cytoplasmic tail of immunoreceptors that is tyrosine phosphorylated by a member of the src tyrosine kinase family, generating docking sites for SH2 containing proteins.

NADPH oxidase—The initiating reaction in the respiratory burst. NADPH oxidase converts molecular oxygen to superoxide which is subsequently dismuted to hydrogen peroxide by superoxide dismutase. These reactive oxygen species have microbicidal activity.

Phagocytosis—Process by which particulate matter is taken up by cells. Professional phagocytes include neutrophils, monocytes, and tissue macrophages.

Phospholipid—Molecule built up on a glycerol backbone. The first two carbons are linked to fatty acids, a saturated fatty acid in the sn-1 position and an unsaturated fatty acid in the sn-2 position. A phosphodiester links the sn-3 carbon with a headgroup, either inositol (phosphatidylinositol, PI), choline (phosphatidylcholine, PC), ethanolamine (phosphatidylethanolamine, PE), or serine (phosphatidylserine, PS). PI and PC are the predominant substrates for generation of second messengers.

Phospholipases—Enzymes that hydrolyze phospholipids. Position of hydrolysis on the gly-cerol backbone defines the class of phospholipase: phospholipase A_2 (PLA_2), phospholipase C (PLC), or phospholipase D (PLD). See Figure 2 for positions of cleavage.

Protein kinase C—Family of serine/threonine kinases that are involved in many signaling networks. The subfamilies are differentiated by their coactivators. Classic PKCs (cPKC, α, β and γ) require a rise in Ca, DAG, and PS for activation, the novel isoforms (δ, ε, η, and θ) are Ca independent but require DAG and PS. PKC-ζ and ι/λ require only PS.

Src homology 2 (SH2) domain—Homologous sequences of ~100 amino acids that recognize phosphorylated tyrosine residues. Found in many molecules, including enzymes, adaptor proteins, and transcription factors.

Acknowledgements

I wish to thank Deborah Moran and Maureen Misuraca for their help in the generation of the figures for this article. Additionally, Drs. Frank Blumenstock, James Drake, and Daniel Loegering for their helpful discussions and suggestions. Dr. Lennartz is supported by NIH (GM-50821).

References

1. Brown GD, Herre J, Williams DL et al. Dectin-1 mediates the biological effects of beta-glucans. J Exp Med 2003; 197(9):1119-1124.
2. Krieger M, Herz J. Structure and functions of multiligand lipoprotein receptors: Macrophage scavenger receptors and LDL receptor-related protein (LRP). Annu Rev Biochem 1994; 63:601-637.
3. Gordon S. Pattern recognition receptors: Doubling up for the innate immune response. Cell 2002; 111(7):927-930.
4. Allen L, Aderem A. Molecular definition of distinct cytoskeletal structures involved in complement- and Fc receptor-mediated phagocytosis in macrophages. J Exp Med 1996; 184:627-637.
5. Vieira OV, Botelho RJ, Grinstein S. Phagosome maturation: Aging gracefully. Biochem J 2002; 366(Pt 3):689-704.
6. Scott CC, Botelho RJ, Grinstein S. Phagosome maturation: A few bugs in the system. J Membr Biol 2003; 193(3):137-152.
7. Amer AO, Swanson MS. A phagosome of one's own: A microbial guide to life in the macrophage. Curr Opin Microbiol 2002; 5(1):56-61.
8. Hackstadt T. The diverse habitats of obligate intracellular parasites. Curr Opin Microbiol 1998; 1(1):82-87.
9. Underhill DM, Ozinsky A. Phagocytosis of microbes: Complexity in action. Annu Rev Immunol 2002; 20:825-852.
10. Kwiatkowska K, Sobota A. Signaling pathways in phagocytosis. Bio Essays 1999; 21:422-431.
11. Garcia-Garcia E, Rosales C. Signal transduction during Fc receptor-mediated phagocytosis. J Leukoc Biol 2002; 72(6):1092-1108.
12. Greenberg S. Modular components of phagocytosis. J Leukoc Biol 1999; 66(5):712-717.
13. Cox D, Berg JS, Cammer M et al. Myosin X is a downstream effector of PI(3)K during phagocytosis. Nat Cell Biol 2002; 4(7):469-477.
14. Fukami K. Structure, regulation, and function of phospholipase C isozymes. J Biochem (Tokyo) Mar 2002; 131(3):293-299.
15. Li Z, Jiang H, Xie W et al. Roles of PLC-beta2 and -beta3 and PI3Kgamma in chemoattractant-mediated signal transduction. Science 2000; 287(5455):1046-1049.
16. Wen R, Jou ST, Chen Y et al. Phospholipase C gamma 2 is essential for specific functions of Fc epsilon R and Fc gamma R. J Immunol 2002; 169(12):6743-6767.
17. Kusner DJ, Barton JA, Wen KK et al. Regulation of phospholipase D activity by actin. Actin exerts bidirectional modulation of Mammalian phospholipase D activity in a polymerization-dependent, isoform-specific manner. J Biol Chem 2002; 277(52):50683-50692.
18. Botelho RJ, Teruel M, Dierckman R et al. Localized biphasic changes in phosphatidylinositol-4,5-bisphosphate at sites of phagocytosis. J Cell Biol 2000; 151(7):1353-1368.
19. Rhee SG, Bae YS. Regulation of phosphoinositide-specific phospholipase C isozymes. J Biol Chem 1997; 272(24):15045-15048.

20. Zheng L, Nibbering PH, van Furth R. Stimulation of the intracellular killing of Staphylococcus aureus by human monocytes mediated by Fcγ receptors I and II. Eur J Immunol 1993; 23:2826-2833.

21. Zheng L, Nibbering PH, Zomerdijk TPL et al. Protein tyrosine kinase activity is essential for Fcγ receptor-mediated intracellular killing of Staphylococcus aureus by human monocytes. Infect Immun 1994; 62:4296-4303.

22. Dusi S, Donini M, Bianca VD et al. Tyrosine phosphorylation of phospholipase C-γ2 is involved in the activation of phosphoinositide hydrolysis by Fc receptors in human neutrophils. Biochem Biophys Res Comm 1994; 201:1100-1108.

23. Lennartz MR, Lefkowith JB, Bromley FA et al. IgG-mediated phagocytosis activates a calcium-independent, phosphatidylethanolamine specific phospholipase. J Leuko Biol 1993; 54:389-398.

24. Della Bianca V, Grzeskowiak M, Rossi F. Studies on molecular regulation of phagocytosis and activation of the NADPH oxidase in neutrophils. J Immunol 1990; 144:1411-1417.

25. Larsen EC, DiGennaro JA, Saito N et al. Differential requirement for classic and novel PKC isoforms in respiratory burst and phagocytosis in RAW 264.7 cells. J Immunol 2000; 165:2809-2817.

26. Kimura T, Kihara H, Bhattacharyya S et al. Downstream signaling molecules bind to different phosphorylated immunoreceptor tyrosine-based activation motif (ITAM) peptides of the high affinity IgE receptor. J Biol Chem 1996; 271(44):27962-27968.

27. Ibarrola I, Vossebeld PJ, Homburg CH et al. Influence of tyrosine phosphorylation on protein interaction with FcgammaRIIa. Biochim Biophys Acta 1997; 1357(3):348-358.

28. Rameh LE, Rhee SG, Spokes K et al. Phosphoinositide 3-kinase regulates phospholipase C-gamma-mediated calcium signaling. J Biol Chem 1998; 273(37):23750-23757.

29. Newton AC. Regulation of protein kinase C. Curr Opin Cell Biol 1997; 9:161-167.

30. Loegering DJ, Lennartz MR. Signaling pathways for Fcγ receptor-stimulated tumor necrosis factor-α secretion and respiratory burst in RAW 264.7 macrophages. Inflammation 2004; 28:23-31.

31. Korchak HM, Rossi MW, Kilpatrick LE. Selective role for beta-protein kinase C in signaling for O-2 generation but not degranulation or adherence in differentiated HL60 cells. J Biol Chem 1998; 273(42):27292-27299.

32. Allen L-AH, Aderem A. A role for MARCKS, the α isozyme of protein kinase C and myosin I in zymosan phagocytosis by macrophages. J Exp Med 1995; 182:829-840.

33. Reeves EP, Dekker LV, Forbes LV et al. Direct interaction between p47phox and protein kinase C: Evidence for targeting of protein kinase C by p47phox in neutrophils [published erratum appears in Biochem J 2000; 345(Pt 3):767]. Biochem J 1999; 344(Pt 3):859-866.

34. Babior BM, Lambeth JD, Nauseef W. The neutrophil NADPH oxidase. Arch Biochem Biophys 2002; 397(2):342-344.

35. DeLeo FR, Allen LA, Apicella M et al. NADPH oxidase activation and assembly during phagocytosis. J Immunol 1999; 163(12):6732-6740.

36. Zheleznyak A, Brown EJ. Immunoglubulin-mediated phagocytosis by human monocytes requires protein kinase C activation. J Biol Chem 1992; 267:12042-12048.

37. Larsen EC, Ueyama T, Brannock PM et al. A role for PKC-varepsilon in FcgammaR-mediated phagocytosis by RAW 264.7 cells. J Cell Biol 2002; 159(6):939-944.

38. Banno Y. Regulation and possible role of mammalian phospholipase D in cellular functions. J Biochem (Tokyo) 2002; 131(3):301-306.

39. Brindley DN, English D, Pilquil C et al. Lipid phosphate phosphatases regulate signal transduction through glycerolipids and sphingolipids. Biochim Biophys Acta 2002; 1582(1-3):33-44.

40. Exton JH. Phospholipase D: Enzymology, mechanisms of regulation, and function. Physiol Rev 1997; 77(2):303-320.

41. Fallman M, Lew DP, Stendahl O et al. Receptor-mediated phagocytosis in human neutrophils is associated with increased formation of inositol phosphates and diacylglycerol. Elevation in cytosolic free calcium and formation of inositol phosphates can be dissociated from accumulation of diacylglycerol. J Clin Invest 1989; 84(3):886-891.

42. Fallman M, Andersson R, Andersson T. Signaling properties of CR3 (CD11b/CD18) and (CD35) in relation to phagocytosis of complement-opsonized particles. J Immunol 1993; 151:330-338.

43. Sciorra VA, Morris AJ. Sequential actions of phospholipase D and phosphatidic acid phosphohydrolase 2b generate diglyceride in mammalian cells. Mol Biol Cell 1999; 10(11):3863-3876.

44. Kusner DJ, Hall CF, Schlesinger LS. Activation of phospholipase D is tightly coupled to the phagocytosis of Mycobacterium tuberculosis or opsonized zymosan by human macrophages. J Exp Med 1996; 184(2):585-595.

45. Kusner DJ, Hall CF, Jackson S. Fcgamma receptor-mediated activation of phospholipase D regulates macrophage phagocytosis of IgG-opsonized particles. J Immunol 1999; 162(4):2266-2274.

46. Anes E, Kuhnel MP, Bos E et al. Selected lipids activate phagosome actin assembly and maturation resulting in killing of pathogenic mycobacteria. Nat Cell Biol 2003; 5(9):793-802.

47. Zhang Q, Cox D, Tseng CC et al. A requirement for ARF6 in fcgamma receptor-mediated phagocytosis in macrophages. J Biol Chem 1998; 273(32):19977-19981.

48. O'Luanaigh N, Pardo R, Fensome A et al. Continual production of phosphatidic acid by phospholipase D is essential for antigen-stimulated membrane ruffling in cultured mast cells. Mol Biol Cell 2002; 13(10):3730-3746.

49. Choi WS, Kim YM, Combs C et al. Phospholipases D1 and D2 regulate different phases of exocytosis in mast cells. J Immunol 2002; 168(11):5682-5689.

50. Di Virgilio F, Meyer CB, Greenberg S et al. Fc Receptor-mediated phagocytosis occurs in macrophages at exceedingly low cytosolic Ca^{+2} levels. J Cell Biol 1988; 106:657-666.

51. Lennartz MR, Yuen AFC, McKenzie Massi S et al. Phospholipase A_2 inhibition results in sequestration of plasma membrane into electron lucent vesicles druing IgG-mediated phagocytosis. J Cell Sci 1997; 110:2041-2052.

52. Lennartz MR. Phospholipases and phagocytosis: The role of phospholipid-derived second messengers in phagocytosis. Int J Biochem Cell Biol 1999; 31(3-4):415-430.

53. Cox D, Tseng CC, Bjekic G et al. A requirement for phosphatidylinositol 3-kinase in pseudopod extension. J Biol Chem 1999; 274(3):1240-1247.

54. Bajno L, Peng XR, Schreiber AD et al. Focal exocytosis of VAMP3-containing vesicles at sites of phagosome formation. J Cell Biol 2000; 149(3):697-706.

55. Katsumata O, Hara-Yokoyama M, Sautes-Fridman C et al. Association of Fc-gammaRII with low-density detergent-resistant membranes is important for cross-linking-dependent initiation of the tyrosine phosphorylation pathway and superoxide generation. J Immunol 2001; 167(10):5814-5823.

56. Castrillo A, Pennington DJ, Otto F et al. Protein kinase C-epsilon is required for macrophage activation and defense against bacterial infection. J Exp Med 2001; 194(9):1231-1242.

57. Shirai Y, Kashiwagi K, Sakai N et al. Phospholipase A(2) and its products are involved in the purinergic receptor-mediated translocation of protein kinase C in CHO-K1 cells. J Cell Sci 2000; 113(Pt 8):1335-1343.

58. Prekeris R, Mayhew MW, Cooper JB et al. Identification and localization of an actin-binding motif that is unique to the epsilon isoform of protein kinase C and participates in the regulation of synaptic function. J Cell Biol 1996; 132:77-90.

59. Prekeris R, Hernandez RM, Mayhew MW et al. Molecular analysis of the interactions between protein kinase C-epsilon and filamentous actin. J Biol Chem 1998; 273(41):26790-26798.

60. Balsinde J, Winstead MV, Dennis EA. Phospholipase A(2) regulation of arachidonic acid mobilization. FEBS Lett Oct 30 2002; 531(1):2-6.

61. Murakami M, Kudo I. Phospholipase A2. J Biochem (Tokyo) 2002; 131(3):285-292.

62. Lennartz MR, Brown EJ. Arachidonic acid is essential for IgG Fc receptor-mediated phagocytosis by human monocytes. J Immunol 1991; 147:621-626.

63. Brown WJ, Chambers K, Doody A. Phospholipase A2 (PLA2) enzymes in membrane trafficking: Mediators of membrane shape and function. Traffic 2003; 4(4):214-221.

64. Mayorga LS, Colombo MI, Lennartz MR et al. A phospholipase A_2 is necessary for endosome fusion. Proc Natl Acad Sci 1993; 90:10255-10259.

65. Bette-Bobillo P, Vidal M. Characterization of phospholipase A_2 activity in reticulocyte endocytic vesicles. FEBS Lett 1995; 228:199-205.

66. Ropert C, Almeida IC, Closel M et al. Requirement of mitogen-activated protein kinases and I kappa B phosphorylation for induction of proinflammatory cytokines synthesis by macrophages indicates functional similarity of receptors triggered by glycosylphosphatidylinositol anchors from parasitic protozoa and bacterial lipopolysaccharide. J Immunol 2001; 166(5):3423-3431.

67. Takuma T, Ichida T. Role of Ca2+-independent phospholipase A2 in exocytosis of amylase from parotid acinar cells. J Biochem (Tokyo) 1997; 121(6):1018-1024.

68. Blackwood RA, Transue AT, Harsh DM et al. PLA_2 promotes fusion between PMN-specific granules and complex liposomes. J Leuk Biol 1996; 59:663-670.

69. Nishio H, Takeuchi T, Hata F et al. Ca(2+)-independent fusion of synaptic vesicles with phospholipase A2- treated presynaptic membranes in vitro. Biochem J 1996; 318(Pt 3):981-987.

70. Karimi K, Lennartz MR. Protein kinase C activation precedes arachidonic acid release during IgG-mediated phagocytosis. J Immunol 1995; 155:5786-5794.
71. Portilla D, Dai G. Purification of a novel calcium-independent phospholipase A$_2$ from rabbit kidney. J Biol Chem 1996; 271:15451-15457.
72. Akiba S, Mizunaga S, Kume K et al. Involvement of group VI Ca2+-independent phospholipase A2 in protein kinase C-dependent arachidonic acid liberation in zymosan-stimulated macrophage-like P388D1 cells. J Biol Chem 1999; 274(28):19906-19912.
73. Balboa MA, Saez Y, Balsinde J. Calcium-independent phospholipase a(2) is required for lysozyme secretion in U937 promonocytes. J Immunol 2003; 170(10):5276-5280.
74. Balboa MA, Shirai Y, Gaietta G et al. Localization of group V phospholipase A2 in caveolin-enriched granules in activated P388D1 macrophage-like cells. J Biol Chem 2003.
75. Gijon MA, Spencer DM, Siddiqi AR et al. Cytosolic phospholipase A2 is required for macrophage arachidonic acid release by agonists that do and do not mobilize calcium. Novel role of mitogen-activated protein kinase pathways in cytosolic phospholipase A2 regulation. J Biol Chem 2000; 275(26):20146-20156.
76. Gijon MA, Leslie CC. Regulation of arachidonic acid release and cytosolic phospholipase A2 activation. J Leukoc Biol 1999; 65(3):330-336.
77. Hazan I, Dana R, Granot Y et al. Cytosolic phospholipase A2 and its mode of activation in human neutrophils by opsonized zymosan. Correlation between 42/44 kDa mitogen- activated protein kinase, cytosolic phospholipase A2 and NADPH oxidase. Biochem J 1997; 326(Pt 3):867-876.
78. Henderson LM, Chappel JB. NADPH oxidase of neutrophils. Biochim Biophys Acta 1996; 1273(2):87-107.
79. Henderson LM, Chappell JB, Jones OT. Superoxide generation is inhibited by phospholipase A2 inhibitors. Role for phospholipase A2 in the activation of the NADPH oxidase. Biochem J 1989; 264(1):249-255.
80. Dana R, Malech HL, Levy R. The requirement for phospholipase A$_2$ for activation of the assembled NADPH oxidase in human neutrophils. Biochem J 1994; 297:217-223.
81. Dana R, Leto TL, Malech HL et al. Levy R. Essential requirement of cytosolic phospholipase A2 for activation of the phagocyte NADPH oxidase. J Biol Chem 1998; 273(1):441-445.
82. Shmelzer Z, Haddad N, Admon E et al. Unique targeting of cytosolic phospholipase A2 to plasma membranes mediated by the NADPH oxidase in phagocytes. J Cell Biol 2003; 162(4):683-692.
83. Lowenthal A, Levy R. Essential requirement of cytosolic phospholipase A(2) for activation of the H(+) channel in phagocyte-like cells. J Biol Chem 1999; 274(31):21603-21608.
84. Levy R, Lowenthal A, Dana R. Cytosolic phospholipase A2 is required for the activation of the NADPH oxidase associated H+ channel in phagocyte-like cells. Adv Exp Med Biol 2000; 479:125-135.
85. Kim D, Clapham DE. Potassium channels in cardiac cells activated by arachidonic acid and phospholipids. Science 1989; 244:1174-1176.
86. Ordway RW, Walsh Jr JV, JJ S. Arachidonic acid and other fatty acids directly activate potassium channels in smooth muscle cells. Science 1989; 244:1176-1179.
87. Gijon MA, Spencer DM, Kaiser AL et al. Leslie CC. Role of phosphorylation sites and the C2 domain in regulation of cytosolic phospholipase A2. J Cell Biol 1999; 145(6):1219-1232.
88. Aderem AA, Cohen DS, Wright SD et al. Bacterial lipopolysaccharides prime macrophages for enhanced release of arachidonic acid metabolites. J Exp Med 1986; 164:165 179.
89. Karimi K, Lennartz MR. Mitogen-activated protein kinase is activated during IgG-mediated phagocytosis but is not required for target ingestion. Inflammation 1998; 22:67-82.
90. Schillace RV, Scott JD. Organization of kinases, phosphatases, and receptor signaling complexes. J Clin Invest 1999; 103(6):761-765.
91. Seger R, Krebs EG. The MAPK signaling cascade. FASEB J 1995; 9:726-735.
92. Morito T, Oishi K, Yamamoto M et al. Biphasic regulation of Fc-receptor mediated phagocytosis of rabbit alveolar macrophages by surfactant phospholipids. Tohoku J Exp Med 2000; 190(1):15 22.
93. Beningo KA, Wang YL. Fc-receptor-mediated phagocytosis is regulated by mechanical properties of the target. J Cell Sci 2002; 115(Pt 4):849-856.
94. Ridley AJ, Hall A. The small GTP-binding protein rho regulates the assembly of focal adhesions and actin stress fibers in response to growth factors. Cell 1992; 70:389-399.
95. Caron E, Hall A. Identification of two distinct mechanisms of phagocytosis controlled by different Rho GTPases. Science 1998; 282(5394):1717-1721.

96. Lee DJ, Cox D, Li J et al. Rac1 and Cdc42 are required for phagocytosis, but not NF-kappa-B-dependent gene expression, in macrophages challenged with Pseudomonas aeruginosa. J Biol Chem 2000; 275(1):141-146.
97. Goni FM, Alonso A. Sphingomyelinases: Enzymology and membrane activity. FEBS Lett 2002; 531(1):38-46.
98. Venkataraman K, Futerman AH. Ceramide as a second messenger: Sticky solutions to sticky problems. Trends Cell Biol 2000; 10(10):408-412.
99. Bourbon NA, Yun J, Kester M. Ceramide directly activates protein kinase C zeta to regulate a stress-activated protein kinase signaling complex. J Biol Chem 2000; 275(45):35617-35623.
100. Bourbon NA, Yun J, Berkey D et al. Inhibitory actions of ceramide upon PKC-epsilon/ERK interactions. Am J Physiol Cell Physiol 2001; 280(6):C1403-1411.
101. Hinkovska-Galcheva V, Boxer L, Mansfield PJ et al. Enhanced phagocytosis through inhibition of de novo ceramide synthesis. J Biol Chem 2003; 278(2):974-982.
102. Breton A, Descoteaux A. Protein kinase C-alpha participates in FcgammaR-mediated phagocytosis in macrophages. Biochem Biophys Res Commun 2000; 276(2):472-476.
103. Allen LA, Allgood JA. Atypical protein kinase C-zeta is essential for delayed phagocytosis of Helicobacter pylori. Curr Biol 2002; 12(20):1762-1766.
104. Suchard SJ, Mansfield PJ, Boxer LA et al. Mitogen-activated protein kinase activation during IgG-dependent phagocytosis in human neutrophils. J Immunol 1997; 158:4961-4967.
105. Mansfield PJ, Carey SS, Hinkovska-Galcheva V et al. Ceramide inhibition of phospholipase D and its relationship to RhoA and ARF1 translocation in GTP{gamma}S-stimulated polymorphonuclear leukocytes. Blood 2003.
106. Suchard SJ, Hinkovska-Galcheva V, Mansfield PJ et al. Ceramide inhibits IgG-dependent phagocytosis in human polymorphonuclear leukocytes. Blood 1997; 89(6):2139-2147.
107. Serrander L, Fallman M, Stendahl O. Activation of phospholipase D is an early event in integrin-mediated signalling leading to phagocytosis in human neutrophils. Inflamm 1996; 20(4):439-450.
108. Lin L-L, Wartmann M, Lin AY et al. cPLA2 is phosphorylated and activated by MAP kinase. Cell 1993; 72:269-278.
109. Cacace A, Ueffing M, Philipp A et al. PKC epsilon functions as an oncogene by enhancing activation of the Raf kinase. Oncogene 1996; 13:2517- 2526.
110. Banno Y, Takuwa Y, Yamada M et al. Involvement of phospholipase D in insulin-like growth factor-I-induced activation of extracellular signal-regulated kinase, but not phosphoinositide 3-kinase or Akt, in Chinese hamster ovary cells. Biochem J 2003; 369(Pt 2):363-368.
111. Jayadev S, Hayter HL, Andrieu N et al. Phospholipase A2 is necessary for tumor necrosis factor alpha-induced ceramide generation in L929 cells. J Biol Chem 1997; 272(27):17196-17203.
112. Baumruker T, Prieschl EE. Sphingolipids and the regulation of the immune response. Semin Immunol 2002; 14(1):57-63.
113. Kwiatkowska K, Frey J, Sobota A. Phosphorylation of FcgammaRIIA is required for the receptor-induced actin rearrangement and capping: The role of membrane rafts. J Cell Sci 2003; 116(Pt 3):537-550.
114. Grazide S, Maestre N, Veldman RJ et al. Ara-C- and daunorubicin-induced recruitment of Lyn in sphingomyelinase-enriched membrane rafts. FASEB J 2002; 16(12):1685-1687.
115. Simonsen A, Wurmser AE, Emr SD et al. The role of phosphoinositides in membrane transport. Curr Opin Cell Biol 2001; 13(4):485-492.
116. Greenberg S, Grinstein S. Phagocytosis and innate immunity. Curr Opin Immunol 2002; 14(1):136-145.
117. Defacque H, Bos E, Garvalov B et al. Phosphoinositides regulate membrane-dependent actin assembly by latex bead phagosomes. Mol Biol Cell 2002; 13(4):1190-1202.
118. Botelho RJ, Scott CC, Grinstein S. Phosphoinositide involvement in phagocytosis and phagosome maturation. Curr Top Microbiol Immunol 2004; 282:1-30.
119. Brumell JH, Grinstein S. Role of lipid-mediated signal transduction in bacterial internalization. Cell Microbiol 2003; 5(5):287-297.
120. Palicz A, Foubert TR, Jesaitis AJ et al. Phosphatidic acid and diacylglycerol directly activate NADPH oxidase by interacting with enzyme components. J Biol Chem 2000; 276:3090-3097.

CHAPTER 9

Calcium Signaling during Phagocytosis

Alirio J. Melendez

Abstract

Phagocytosis is important for a wide diversity of organisms. From simple unicellular organisms that use phagocytosis to eat, to complex metazoans in which phagocytic cells represent an essential branch of the immune system. Evolution has armed cells with a fantastic repertoire of molecules that serve to bring about this complex event regardless of the organism or specific molecules concerned. However, all phagocytic processes are driven by a finely controlled rearrangement of the actin cytoskeleton where calcium (Ca^{2+}) signals play important roles. Ca^{2+} plays many roles in cytoskeletal changes by affecting the actions of a number of contractile proteins, as well as being a cofactor for the activation of a number of intracellular signaling proteins, known to play important roles during phagocytosis. In the mammalian immune system, the requirement of Ca^{2+} for the initial steps in phagocytosis, and the subsequent phagosome maturation, can be quite different depending on the type of cell and on the type of receptor that is driving phagocytosis.

Introduction

Phagocytosis, the internalization of large particles by cells, was first described by the Russian scientist Elie Metchnikoff in the late eighteen hundreds, nearly 120 year ago.[1] Metchnikoff first described how "amoeboid" cells moved, within a transparent starfish larvae, towards an inserted rose thorn and engulfed the thorn. Metchnikoff named this process "phagocytosis". Today, phagocytosis is used to define the cellular engulfment of particles larger than 0.5 μm in diameter.

Phagocytosis is usually associated with the function of the immune system in relation to the elimination of invading microorganisms, foreign particles, and the elimination of infected or dying cells.[2] However, this process is not restricted to the function of immune cells or indeed to the function of mammalian cells, but developed early in evolution and is present in *Dictyostelium*, nematodes and insect haemocytes.[3,4]

Phagocytosis is important for a wide diversity of organisms. From simple unicellular organisms that use phagocytosis to obtain their next meal, to complex metazoans in which phagocytic cells represent an essential branch of the immune system, evolution has armed cells with a fantastic repertoire of molecules that serve to bring about this complex event. Regardless of the organism or specific molecules concerned, however, all phagocytic processes are driven by a finely controlled rearrangement of the actin cytoskeleton. Calcium plays many roles in cytoskeletal changes by affecting the actions of a number of contractile proteins. It has been shown that calcium can affect cytoskeletal changes by stimulating

Molecular Mechanisms of Phagocytosis, edited by Carlos Rosales. ©2005 Eurekah.com and Springer Science+Business Media.

myosin contractility,[5] activation of actin filament severing,[6] inhibition of actin crosslinking by α-actinin,[7] or the 34kDa actin-crosslinking protein,[8] binding to annexins,[9] calmodulin,[10] or other low molecular weight calcium-binding proteins including CBP1-4,[11] calpain,[12] or other potential targets.

In mammals, several cell types are capable of phagocytosis, but their phagocytic activity is very varied. This fact is reflected by dividing the cells that are capable of phagocytosis into: *professional phagocytes*; *paraprofessional phagocytes*; and *nonprofessional phagocytes,* this terminology refers to cells with high, medium or low phagocytic activity respectively.[13] Because of the large variety and complexity of phagocytic cells, the scope of this chapter will be limited to the *professional phagocytes.*

Professional phagocytes encompass mainly neutrophils and cells of the monocytic/macrophage lineage, sentinels of the immune system that hunt and destroy senescent, apoptotic or otherwise defective host cells, pollutant particles and, perhaps most importantly, foreign, potentially pathogenic organisms.[13,14] The unique ability of phagocytic leukocytes to efficiently internalize a variety of targets is attributable to the expression of an array of specialized phagocytic receptors. Supporting this notion, it has been shown that the phagocytic capacity of nonprofessional phagocytes, such as Chinese hamster ovary or COS cells, is greatly increased by the heterologous expression of specialized phagocytic receptors, such as Fcγ receptors (FcγRs), that are normally found in neutrophils or macrophages.[15]

In mammalian cells, phagocytosis is a receptor-mediated and actin-dependent process. Macrophages and neutrophils eliminate invading pathogens by first ingesting them into a plasma membrane-derived vacuole, named phagosome. The resulting phagosomes undergo a series of fission and fusion events that modify the composition of their limiting membrane and of their contents, by a process termed phagosomal maturation, which empowers the vacuole with a host of degradative properties central to the destruction of the invading pathogen.[16] However, phagocytosis can also have unwanted effects for the host in that certain pathogens, such as *Mycobacterium tuberculosis*, take advantage of the phagocytic machinery to gain access to the cell interior, without being destroyed by the phagocytic machinery, and become intracellular pathogens.[17,18]

Receptors Involved in Phagocytosis

Phagocytosis is initiated by the interaction of surface receptors with their cognate ligand. Ligands can be endogenous components of the particle, exemplified by lipopolysaccharides of bacteria and phosphatidylserine in apoptotic cells.[14,18] Internalization triggered by endogenous ligands of the particle is known as nonopsonic. The immune system is equipped with a variety of receptors that recognize nonopsonic ligands, including CD14 that binds to lipopolysaccharides, as well as receptors that recognize specifically phosphatidylserine, mannose or fucose residues.[19] Alternatively, phagocytic ligands can be classified as opsonins, which are host-derived proteins that coat the surface of a particle. The best characterized opsonins are the complement fragment C3b, iC3b and IgG antibodies. C3b and iC3B bind relatively nonspecifically to the surface of foreign particles, whereas IgG molecules attach to the phagocytic target by recognizing specific surface epitopes.[14] C3b or iC3b-opsonized particles are recognized by complement receptors members of the integrin superfamily, while IgG-opsonized particles engage FcγRs.[15] In any case, receptor engagement leads to internalization of the particle into a phagosome by a complex sequence of events that require kinase activation, alterations in phospholipid metabolism, remodeling of the actin cytoskeleton and acceleration of membrane traffic.[3,14] These processes that are usually associated with increase in intracellular calcium.

In mammals, binding of immunoglobulins (Igs) to foreign particles (opsonization) leads to the prompt clearance of those particles from the organism. The conserved Fc-domains of the Igs are recognized by Fc receptors present on professional phagocytes, such as macrophages and neutrophils. An opsonized particle binds to Fc-receptors and is rapidly internalized by an actin-dependent extension of the plasma membrane around the opsonized particle. This process is accompanied by the production of proinflammatory and toxic molecules, such as the production of superoxide and the release of cytokines from the phagocytic cell.[20] The major Ig opsonin is IgG, which binds to the corresponding Fc-gamma-receptors (FcγRs), although IgA and IgE also have cognate Fc receptors (FcαRs and FcεRs, respectively) that are involved in phagocytosis.[21,22]

A range of FcγRs exist.[23] FcγRI, FcγRIIA, and FcγRIIIA can all support phagocytosis.[24-26] FcγRIIB negatively regulates phagocytosis;[27] while FcγRIIIB is able to initiate calcium signaling and actin polymerization, but its role in phagocytosis remains unclear.[28-30] Fc receptor-mediated phagocytosis is fully reviewed in.[31]

Another group of receptors that mediate phagocytosis by professional phagocytes is the complement receptors. Complement-receptor-mediated phagocytosis is morphologically distinct from that mediated by FcRs, although both processes require actin polymerization. Complement-opsonized particles get internalized, with minimal membrane disturbance and this does not usually lead to an inflammatory response or the generation of superoxide. The complement system is evolutionarily much older than adaptive immunity and is present even in simple organisms such as sea urchins,[32] and yet still represents an important part of the innate immune system in higher organisms, including humans.[33] In higher vertebrates the complement system is composed of at least 30 proteins, which are activated by enzymatic cascades, by exposure to microbial macromolecules or by binding to antibodies (primarily IgM or IgG) bound to the surface of a pathogen. One of the molecules produced following the complement cascade is C3b. This molecule can bind to molecules on microbial surfaces, where C3b acts like an opsonin, and is recognized by the complement receptor 1 (CR1 also known as CD35). C3b can be further modified by plasma factors H and I, which convert it to iC3b. iC3b is a very potent opsonin that can be recognized by the complement receptor 3, CR3, also known as Mac-1, CD11b/CD18 or αMβ2 integrin; iC3b can also be recognized by complement receptor 4, CR4, also known as CD11c/CD18 or αxβ2 integrin. The complement receptors CR1, CR3 and CR4, are expressed on macrophages and neutrophils and are capable of mediating phagocytosis.[34] CR3 has been the most widely studied of the complement receptors and is capable of binding to several ligands through different recognition sites.[35] However, the phagocytosis of iC3b-opsonised particles by CR3 can only proceed efficiently if the phagocytes are first activated, either by proinflammatory cytokines/chemokines, or by binding to the extracellular matrix.[36,37] It is believed that the preactivation triggers a conformational change in the complement receptor,[38] possibly through phosphorylation of the β subunit.[39] This triggers clustering of the receptor;[40] a precondition for particle binding, which allows the phagocytosis to[41] occur.

A Variety of Nonprofessional Phagocytic Receptors

It is becoming apparent that a growing number of cell-surface receptors can mediate phagocytic uptake of particles. These include noncomplement-receptor integrins such as α5β1 and αvβ3, which mediate uptake of particles coated with fibronectin,[42] lectins such as the mannose receptor,[43] the lipopolysaccharide (LPS) receptor CD14,[44] and the diverse scavenger receptor group.[45] Internalization by these receptors appears, at least in some cases, to be morphologically dynamic, as in the case of Fc receptors, but in contrast to uptake through

CR3. The membrane is extended around the attached particle, and there is transient ruffling in surrounding areas of the cell.[46] However, uptake does not trigger inflammation,[47] and might actively suppress it.[48,49] Recently, there has been a resurgence of interest in these receptors in an attempt to understand the removal of apoptotic cells.[45] In metazoans development is accompanied by massive apoptosis (e.g., during limb formation), and these 'corpses' are engulfed both by professional phagocytes and by neighboring cells that act as 'nonprofessional' phagocytes.[50] The recent report of a receptor for phosphatidylserine,[51] a membrane phospholipid exposed externally on apoptotic cells, is likely to stimulate rapid progress in this field. Additionally, *Caenorhabditis elegans* appears to use a common mechanism to engulf apoptotic and necrotic cell corpses,[52] whether this is also true for 'higher' organisms remains to be seen. In common with FcR- and complement-receptor-mediated phagocytosis, phagocytosis mediated by this diverse group of receptors is also actin dependent,[45] and many of the downstream components are the same as those lying downstream of CR3 or FcγRs. Nevertheless, the events immediately following receptor-ligand interaction remain largely unknown. Some of these receptors might act only to tether particles, and then utilize accessory receptors to deliver the phagocytic signal.[50] This would explain the ability of some receptors (such as CD14) to induce inflammatory responses when binding to one ligand (LPS) but not another.[44] Some of these receptors can signal through tyrosine kinases, and uptake is, at least in some cases, phosphorylation dependent.[53] In this regard it is particularly intriguing that the newly cloned *ced-1* gene from *C. elegans* (which encodes a transmembrane receptor that is essential for the uptake of apoptotic cells) contains an intracellular YXXL motif.[54] This sequence is also found in the ITAMs of mammalian FcRs, where it mediates interactions with downstream signaling elements.

Phagocytosis has been known to be an actin-dependent process since 1977, when it was reported that cytochalasin B, a toxin that blocks actin polymerization, inhibited the uptake of IgG-coated erythrocytes by mouse macrophages.[55] Although originally proposed to be a unique feature of Fc-receptor mediated phagocytosis, remodeling of the actin cytoskeleton is now known to be required for phagocytosis through other types of receptor as well. However, one key characteristic for all receptor-triggered actin remodeling is the sensitivity to changes in intracellular calcium.

Calcium the Ubiquitous Second Messenger

Cytosolic Ca^{2+} is a ubiquitous intracellular signal, pivotal in many signal transduction pathways, controlling a wide range and diversity of cellular activities ranging from proliferation and differentiation to cell death.[56] Resting cells have a Ca^{2+} concentration around 100 nM, this amount of intracellular Ca^{2+} is not sufficient to trigger substantial cellular activities, however, when cells are stimulated the amount of intracellular Ca^{2+} can rise very quickly reaching up to 1 µM, and Ca^{2+} triggered cellular activities occur.[56] It is well established that the increase of cytosolic Ca^{2+} can be temporally and spatially very complex. This is due to the fact that different cells respond differently to a particular stimulus, and indeed different stimuli triggered a particular cell in different ways. Thus, the Ca^{2+} signals can of a single burst and very transient or long-lasting and oscillatory, and can happen in a localized microenvironment or can be triggered as a wide-spread event.[57]

Cells generate their Ca^{2+} signals by using both internal and external sources of Ca^{2+}. Internally Ca^{2+} is stored in specialized compartments such as the endoplasmic reticulum (ER), sarcoplasmic reticulum (SR), or on smaller compartments called calciosomes which are thought to be present in many cell types.[58] Endosomes and phagosomes also store calcium.[59] Ca^{2+} signals are controlled by the generation of intracellular second messengers binding to specific receptors/channels. There are several intracellular second messengers known to increase cytosolic Ca^{2+}, these include: inositol 1,4,5-trisphosphate (IP_3); cyclic adenosine 5'-diphosphoribose (cADPR); nitric oxide (NO); hydrogen peroxide (H_2O_2); superoxide (O2-); nicotinic acid adenine dinucleotide phosphate (NAADP); diacylglycerol (DAG); arachidonic acid (AA); phosphatidic acid (PA); sphingosine; sphingosine-1-phosphate (S1P); and Ca^{2+} itself.[57] Of these intracellular second messengers, some act on intracellular Ca^{2+} channels found on internal compartments for Ca^{2+} release, some act on Ca^{2+} entry channels found on the plasma membrane, while some may act on both release and entry.[56,57] Due to the specificity of these Ca^{2+} triggering messengers we can say that there must be several different types of intracellular Ca^{2+} release channels, however, only IP_3 receptors and ryanodine receptors (RYRs) have been well characterized. Therefore, Ca^{2+}-mobilizing second messengers, generated when cell-surface receptors are stimulated, determine whether Ca^{2+} channels can be activated. Thus, generated IP_3 can diffuse in the cell's cytoplasm to engage the IP_3 receptors and release Ca^{2+} from the ER.[56] The activity of the RYRs is modulated by the generation of cADPR,[58] NAADP acts on a separate-uncharacterized channel.[59] Exactly how sphingosine-1-phosphate causes Ca^{2+} release from intracellular stores is still unclear. Until recently, the best candidate for the sphingosine-1-phosphate receptor was a protein known as 'sphingolipid Ca^{2+}-release mediating protein of endoplasmic reticulum' (SCaMPER).[60] It had been proposed that this protein formed a widely occurring channel responsive to sphingosine-1-phosphate and sphingosylphosphorylcholine. A recent reinvestigation of SCaMPER found that there was little correlation between its intracellular location and that of known intracellular Ca^{2+} stores.[61] Furthermore, expression of SCaMPER was found not to confer sensitivity to sphingolipids, nor to affect Ca^{2+} homeostasis, but could lead to cell death.

A range of Ca^{2+} influx channels have been established for some time. In addition to their activation by some of the messengers listed above, Ca^{2+} influx channels are activated by stimuli including membrane depolarization, stretch, noxious stimuli, extracellular agonists and depletion of intracellular stores.[56,57] Recent studies have expanded the numbers of Ca^{2+}-increasing messengers and channels yet further. In addition, it is becoming evident that different Ca^{2+} signaling pathways can interact to control the source and characteristics of cytosolic Ca^{2+} signals.[56,57] Depletion of intracellular Ca^{2+} stores by activation of IP_3 or ryanodine receptors switches on a Ca^{2+} influx pathway through storeoperated channels (SOCs).[62] The mechanism underlying SOC activation, and the identity of the channels involved, is unclear. It was thought that SOCs provide the main route for Ca^{2+} entry into nonelectrically excitable cells. However, accumulating evidence suggests that intracellular messengers can activate Ca^{2+} influx during physiological stimulation. When IP_3 is produced from phosphoinositide hydrolysis, there is a concomitant production of diacylglycerol (DAG). Unlike the water soluble IP_3, DAG stays in the plane of the plasma membrane where it can activate protein kinase C (PKC) or be metabolized in various ways. PKC and DAG have both been shown to cause Ca^{2+} influx distinct from SOC.[63] Furthermore, other messengers resulting from DAG metabolism, including arachidonic acid and leukotrienes, activate nonSOC (NSOC) Ca^{2+} influx.[64] At present, the best molecular candidates for SOC and NSOC channels are the TRP proteins (so-called because of their homology with the transient receptor potential protein that underlies phototransduction in *Drosophila*). The TRP superfamily has been subdivided into multiple subfamilies on the basis of sequence similarity.[65] In the case of SOC, much attention has been focused on the canonical TRP (TRPC) subfamily. Despite considerable effort, it is unclear exactly which of the seven TRPC isoforms are the molecular constituents of endogenous SOCs. It was recently suggested that a channel belonging to the TRPV subfamily, CaT1 (TRPV6), could be a candidate for a form of SOC known as ICRAC (Ca^{2+}-release activated current).[66] However, the correlation between CaT1

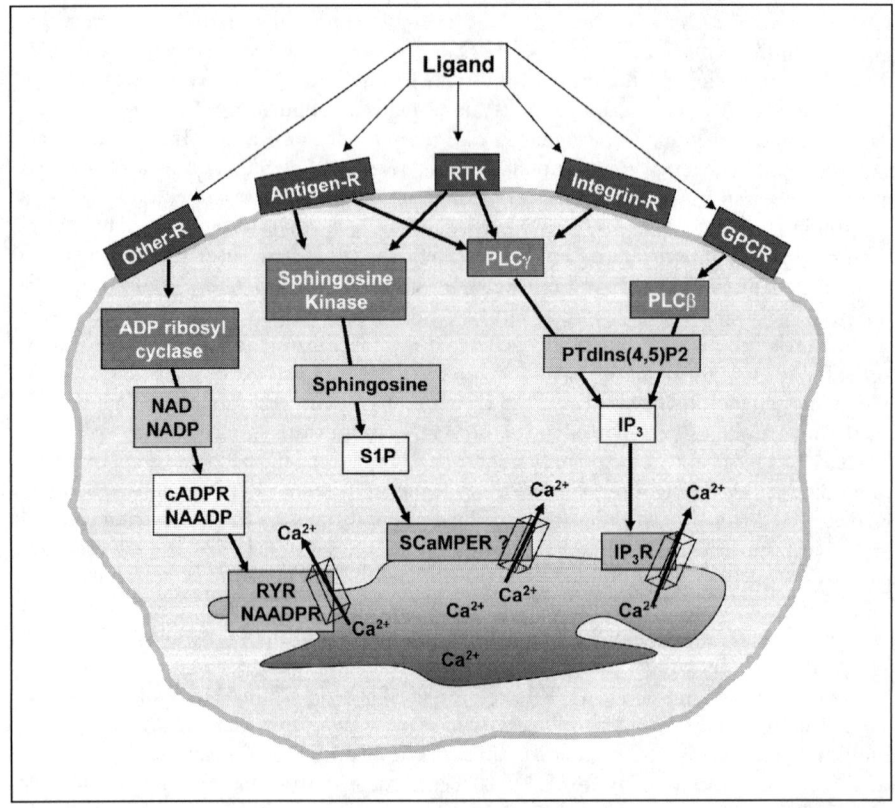

Figure 1. Diagram showing the various signals that trigger calcium-release from internal stores. Mobilization of Ca^{2+} from intracellular stores is achieved by a number of signals generated by ligands binding to a variety of cell-surface receptors, including phagocytic receptors (Antigen-R), receptor tyrosine kinase (RTK), integrin receptors (integrin-R), G-protein coupled receptors (GPCR), and other type of receptors (Other-R) capable of triggering Ca^{2+} release from internal stores. The signals generated include: cyclic ADP ribose (cADPR) and nicotinic acid dinucleotide phosphate (NAADP), both generated from nicotinamide-adenine dinucleotide (NAD) and its phosphorylated derivative NADP by ADP ribosyl cyclase; sphingosine 1-phosphate (S1P), generated from sphingosine by a sphingosine kinase; and the most classical one, inositol-1,4,5-trisphosphate (IP$_3$), generated by the hydrolysis of phosphatidylinositol-4,5-bisphosphate (PtdIns(4,5)P$_2$) by a family of phospholipase C enzymes (PLCγ, PLCβ). The intracellular Ca^{2+} channels: the IP$_3$ receptor (IP$_3$R), ryanodine and NAADP receptors (RYR, NAADPR), and the putative sphingolipid Ca^{2+} release-mediating protein of the ER (SCaMPER).

and ICRAC has been disputed.[67] (Fig. 1) shows a general reprensetation of receptor-triggered calcium release from internal stores.

What Couples Phagocytic-Receptor Activation to Rise on Intracellular Ca^{2+}

Ca^{2+} is a key second messenger in leukocyte activation. It mediates, at least in part, activation of the respiratory burst and secretion of microbicidal granule constituents.[68,69] As in other systems, the resting cytosolic free Ca^{2+} concentration (intracellular Ca^{2+}) hovers in the 100 nM range, but is acutely elevated upon the engagement of phagocytic receptors.[70,71] Release of

Ca^{2+} stored in the endoplasmic reticulum and opening of storeoperated channels are largely responsible for this elevation.

It has been realized recently that organelles other than the endoplasmic reticulum can contribute to the elevation of intracellular Ca^{2+}. Ca^{2+} is now thought to be released also by early and late endosomes, lysosomes and the yeast vacuole.[72,73] Along the same lines, it is entirely conceivable that Ca^{2+} trapped in the lumen of forming phagosomes, or accumulated afterwards by plasmalemmal Ca^{2+} pumps, may be released at critical stages of the maturation sequence. Indeed, preliminary evidence to this effect has been presented.[74] Consistent with this model, a localized periphagosomal increase in intracellular Ca^{2+} has been recorded,[70] although this was attributed to the preferential distribution of endoplasmic reticulum in the immediate vicinity of phagosomes.

The release of Ca^{2+} from internal stores, following receptor engagement in immune cells, is triggered by phospholipid-derived second messengers. Inositol-1,4,5-trisphosphage (IP_3) is the best characterized second messenger responsible for triggering Ca^{2+} release from internal stores.[75] However, Fc-receptor-triggered Ca^{2+} release from internal stores in neutrophils, mast cells and monocytes has also been shown to be IP_3 independent.[76-81] Indirect evidence suggests that L-plasmin, an actin-binding protein, phosphorylated in response to phagocytosis, might participate in the IP_3-independent Ca^{2+} increase mediated by FcγRIIA in neutrophils.[82] Furthermore, it has been shown that the second messenger sphingosine-1-phosphate is the actual trigger responsible for the release of Ca^{2+} from internal stores stimulated by FcγRI aggregation in monocytes or FcεRI aggregation in mast cells.[77-81] However, when monocytes are differentiated to a more macrophage phenotype, FcγRI triggers PLCγ activation and Ca^{2+} signals that are IP_3-dependent.[78] Of interest, is has recently been reported that in human mast cells, FcεRI triggers a dual calcium response.[81] Thus, mast cells appear to concurrently utilize different messengers; in this case IP_3 and sphingosine-1-phosphate to generate the Ca^{2+} signals that underlie the synthesis and release of inflammatory mediators.[81] Essentially, the FcεRI, antigen receptor, on these cells trigger multiple signaling pathways. One of these is phosphoinositide hydrolysis, leading to IP_3 production. Another is the stimulation of phospholipase D, which hydrolyses phosphatidylcholine into phosphatidic acid and choline. It has been suggested that phosphatidic acid can activate a kinase that phosphorylates sphingosine into sphingosine-1-phosphate.[83] The dual activation of these pathways leads to a Ca^{2+} signal with a rapid peak (sphingosine-1-phosphate dependent) and a sustained plateau (IP_3 dependent).

Downstream Events Triggered by Ca^{2+} following Phagocytic-Receptor Activation

One of the first reported signals to be observed in response to phagocytic-receptor activation is an increase in intracellular Ca^{2+} concentration.[84] In neutrophils, this calcium pulse has been reported to be required for FcγR-mediated phagocytosis by one group,[85] but not by another.[86] In contrast, CR3-mediated phagocytosis in neutrophils appears to be independent of changes in intracellular Ca^{2+} concentration, at least in the initial stages of phagocytosis.[86,87] However, following the internalization of IgG-opsonized particles by neutrophils, Ca^{2+} appears to trigger actin depolymerization at phagosomes,[88] a step that may be necessary for phagosome-lysosome fusion.[89] In macrophages, there is a degree of controversy on the role of Ca^{2+} during phagocytosis; some groups report that neither phagocytosis nor phagosome-lysosome fusion are calcium-dependent,[90,91] proposed to reflect the involvement of different phagocytic receptors;[92] whereas, in other studies it has been shown that during phagocytosis by macrophages, there is a rise in Ca^{2+} concentration in the cytoplasm surrounding the phagocytic cup,[93] it is believed that this rise in intracellular Ca^{2+} is directly triggered by receptor activation during phagocytosis, and that it is important at least for phagosome maturation.[93,94,95] However, some

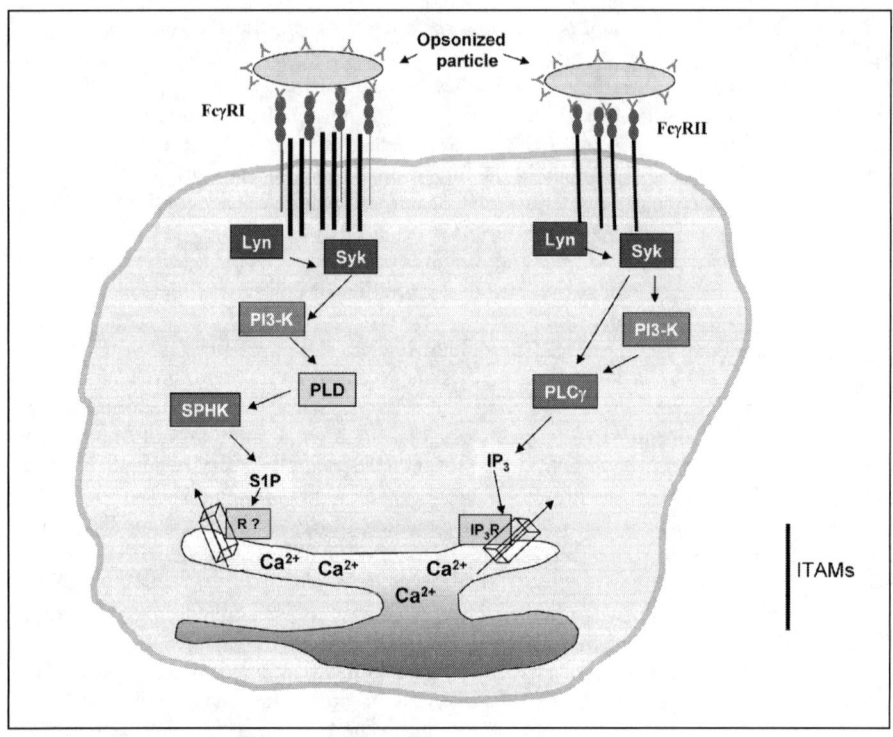

Figure 2. Diagram showing two different pathways utilized by FcγRs to release Ca^{2+} from internal stores during a phagocytosis. IgG-opsonized particles binding to FcγRI or FcγRIIa stimulate Src family kinases, which phosphorylate key tyrosine residues in the ITAMs. The phosphorylated ITAMS serve as docking sites for the tyrosine kinase Syk, and Syk in turn phosphorylates the p85-subunit of phosphatodylinositol-3-kinase (PI3-K). Following these steps the two pathways may divert, depending on cell type and or/stage of phagocytic cell maturation. Thus, in the FcγRI pathways phospholipase D (PLD) is activated upstream of sphingosine kinase (SPHK), and SPHK generates sphingosine-1-phosphate (S1P), which in turns triggers the release of calcium from internal stores, possibly by binding to the SCaMPER or another receptor-channel. On the other hand, FcγRIIa signaling pathways follows the more conventional route, that is, the activation of phospholipase C-gamma (PLCγ). PLCγ hydrolyses of phosphatidylinositol-4,5-bisphosphate ($PtdIns(4,5)P_2$), to generate inositol-1,4,5-trisphosphate (IP_3). IP_3 releases Ca^{2+} from intracellular stores by binding to the IP_3receptors.

studies show that this increase in intracellular Ca^{2+} may be caused by the exit of calcium from the phagosome through Ca^{2+} channels, rather than by Ca^{2+} release from intracellular stores.[94] In this case the reduction of Ca^{2+} concentration in the phagosome seems to be important for phagosome maturation.[94] Independently of its origin, Ca^{2+} seems to be important for triggering depolymerazition around the phagosomes.[95]

How Might Intracellular Ca^{2+} Control Actin Depolymerization during Phagocytosis?

One possible model is that a local rise in intracellular Ca^{2+} concentration activates gelsolin. Gelsolin caps the barbed (fast growing) end of actin filaments, preventing filament elongation, and can also sever filaments in a Ca^{2+}-dependent manner.[96,97] Gelsolin localizes to nascent phagosomes in macrophages.[97] Furthermore, neutrophils from gelsolin knockout (Gsn$^{-/-}$) mice have a profound defect in Fcγ R-mediated phagocytosis.[98] However, the Ca^{2+}-dependent depolymerization of actin filaments from around particles after internalization is normal in Gsn$^{-/-}$ neutrophils,[98] which suggest that intracellular Ca^{2+} plays a role wider than simply activating gelsolin. However, there are other powerful pathways, important for phagocytosis, that require Ca^{2+} signals to be activated, such as is the case with the protein Kinase C family of serine/threonine kinases.

The PKC family of serine/threonine kinases, are activated by the phospholipase product diacylglycerol (DAG), by Ca^{2+}, and by pharmacological agents such as phorbol esters. PKCα localizes to macrophage phagosomes during FcγR-, CR3- and mannose-receptor-mediated phagocytosis.[99,100,101] Complement-receptor-mediated phagocytosis, both of iC3b and β-glucan-opsonized particles, appears to require PKC activity.[53,99] In contrast, conflicting results have been obtained for its involvement in FcγR-mediated phagocytosis.[99,101-103] In addition to the PKCα isoform, PKCβ,[104] PKCγ,[105] PKCδ and PKCε,[101] have all been shown to localize to the phagosome membrane during FcγR-mediated phagocytosis. The isoforms recruited may depend on the differentiation state of the cells and/or the exact FcγR involved,[105] or different isoforms controlling different aspects of phagocytosis.[101] PKC has a range of downstream targets that are implicated in phagocytosis. For example, plekstrin, the major PKC phosphorylation target in platelets, is expressed in macrophages and recruited to the phagosome membrane during FcγR-mediated phagocytosis,[106] although its role there is unknown. More is known about MARCKS and MacMARCKS, two other PKC targets implicated in phagocytosis. MARCKS (myristoylated alanine-rich C kinase substrate) and the closely related MacMARCKS,[107] are actin-filament crosslinking proteins that can also link actin filaments to the membrane.[108] MARCKS localizes to phagosomes,[99,100] and becomes phosphorylated during zymosan phagocytosis.[100] However, macrophages derived from MARCKS$^{-/-}$ mice show normal rates of FcγR- and CR3-mediated phagocytosis and only a minor reduction in the uptake of zymosan particles.[109] MacMARCKS also localizes to phagosomes.[110,111] Mutations in MacMARCKS were reported to block zymosan phagocytosis (a mannose-receptor-mediated event) by one group,[111] but MacMARCKS$^{-/-}$ macrophages do not show phagocytic defects, a discrepancy that might be attributable to the use of different cell lines.[110]

Role of Ca^{2+} on the Maturation of Phagosomes during Phagocytosis?

Release of Ca^{2+} from intracellular stores could have significant implications for membrane fusion, in that a localized amount of high intracellular Ca^{2+} may form in the immediate vicinity of a phagosome, promoting and targeting fusion with cognate vesicles. In fact, a number of studies implicate intraorganellar Ca^{2+} in the homotypic fusion of early endosomes and yeast vacuoles, and in late-endosome–lysosome heterotypic fusion.[72,73,112] In the endocytic pathway, the effects of the intracellular Ca^{2+} released locally are thought to be mediated by calmodulin,[72,73,112] which it is proposed acts downstream of the Rab GTPases and SNARE complex by promoting bilayer coalescence.[73,113]

While the evidence implicating Ca^{2+} in the endocytic pathway is well reported, the role of Ca^{2+} in phagosome maturation is far from clear. There is at least one convincing report that discounts a role for intracellular Ca^{2+} in phagolysosome biogenesis in macrophages.[114] By contrast, preventing changes in intracellular Ca^{2+} was shown to impair phagosome–lysosome fusion in neutrophils and macrophages.[115] Similar observations were also obtained with engineered phagocytes expressing FcγRIIA.[116] Moreover, clamping intracellular Ca^{2+} prevented efficient killing of *Staphylococcus* by neutrophils, suggesting that phagosome maturation was defective.[117]

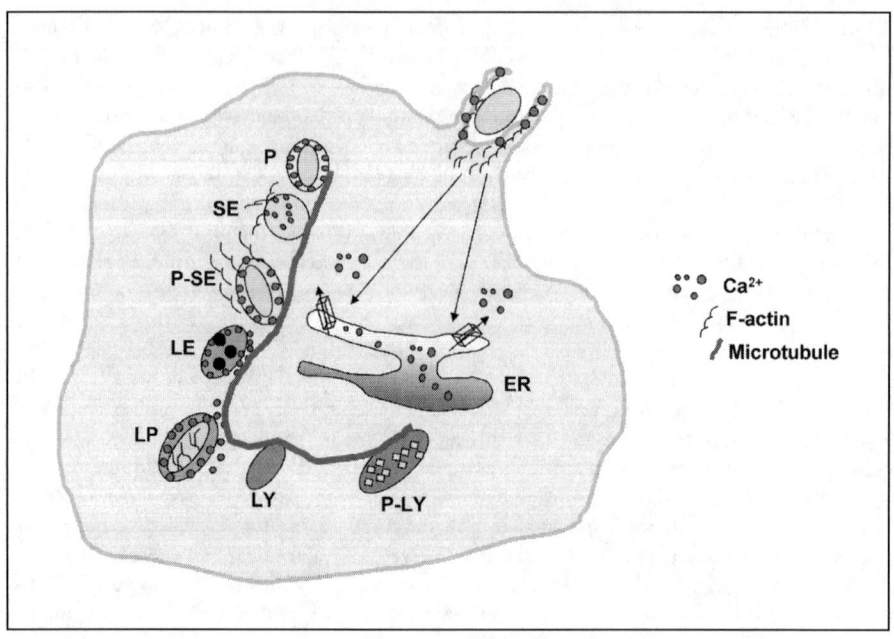

Figure 3. Diagram illustrating phagosome maturation to phagolysosome and the potential role for Ca^{2+} in the process. The initial step in phagosome maturation is believed to be the fusion of the sealed phagosome (P) with sorting endosomes (SE), crating a phago-sorting endosome (P-SE). Following this the P-SE fuses with late endosomes (LE) forming a late phagosome (LP). The late phagosome fuses with the lysosomes (LY) generating the phagolysososme (PLY) where full degradation occurs. Ca^{2+} can be triggered from the endoplasmic reticulum (ER) by the initial phagocytic-receptor activation. Trapping of Ca^{2+} during phagocytosis and the possibility of regulated organellar Ca^{2+} release into the cytosol are illustrated. Note that endosomes, lysosomes and phagosomes at various stages of maturation are thought to bind to microtubules and can serve to nucleate actin. Actin is also abundant at the phagocytic cup, and actin-dependent depolymerization requires Ca^{2+}.

However, how intracellular Ca^{2+} may regulate phagosome maturation is still not well understood. One possibility is that intracellular Ca^{2+} induces disassembly of the actin coating the surface of the phagosome, permitting access to incoming vesicles.[95] Of interest, it has been reported that phagosome-endosome fusion is impaired when periphagosomal actin is stabilized.[118] Interestingly, retention of an actin coat around *Mycobacterium*-containing phagosomes by inhibition of Ca^{2+} is consistent with the presence of coronin, an actin-binding protein, in *Mycobacterium*-containing phagosomes.[119] Nonetheless, exceptions have been reported: in macrophages, actin assembly and disassembly appeared to be normal when Ca^{2+} was clamped at a very low concentration[120] Alternatively, Ca^{2+} may regulate fusion by a more direct approach, through annexins, calmodulin and/or Ca^{2+}/calmodulin-dependent protein kinase [73,112,121-123] Calmodulin and the calmodulin-dependent kinase may in turn regulate tethering or docking factors such as EEA1 and syntaxin 13, and/or regulate bilayer fusion between phagosomes and endo/lysosomes.[73,113]

Conclusion

Despite the great amount of data acquired since Metchnikoff's initial description of the phagocytic process, still we remain with enormous gaps in our understanding of phagocytosis, and can only now realize the complexity of the process and the many issues that need to be addressed. Notable among these are the determinants or selectivity that dictates particle uptake and couples this process with the progressive fusion and elements of the endocytic pathway. It is important to notice the role of Ca^{2+} and the cytoskeleton in the uptake and the process of phagosome maturation. Although in the past few years the amount of data on the role of Ca^{2+} on phagocytosis has exponentially increased, there is still some controversy about the role(s) played by Ca^{2+} during phagocytosis: the relevance of Ca^{2+} for phagocytosis is challenged by the fact that some studies have shown that some receptors may not utilize Ca^{2+} for the initial stages of phagocytosis, such as FcγRI-mediated phagocytosis; however, other receptors such as FcγRII-mediate phagocytosis in a Ca^{2+}-dependent manner. Despite the controversy, a role for Ca^{2+} is becoming well established in endosome-phagosome fusion, phagosome maturation, and in the general actin-cytoskeleton remodeling during phagocytosis. Thus, whereas Ca^{2+} may not be a prerequisite for the initial particle uptake for some receptors, it is becoming clearer that Ca^{2+} plays a major role in phagosome maturation and potentially in antigen presentation.

The Ca^{2+} signaling field is moving forward in many areas: we now have a clearer picture of how different messengers and channels generate Ca^{2+} signals, and their roles in different cellular processes. These advances are coupled with the significant progress made in identifying molecules that are potentially involved in phagocytic uptake, and phagosome maturation. An important challenge for the future is to discover how these mechanisms relate to each other, and how Ca^{2+} signals triggered by different receptor-systems may regulate the process of phagocytosis.

References

1. Metchnikoff E. Sur la lutte des cellules de l'organisme contre l'invasion des microbes. Ann Inst Pasteur 1887; 1:321-345.
2. Aderem A, Underhill DM. Mechanisms of phagocytosis in macrophages. Annu Rev Immunol 1999; 17:593-623.
3. Cardelli J. Phagocytosis and micropinocytosis in Dictyostelium: Phosphoinositidebased processes, biochemically distinct. Traffick 2001; 2:311-320.
4. Franc NC, White K, Ezekowitz RA. Phagocytosis and development: Back to the future. Curr Opin Immunol 1999; 11:47-52.
5. Tan Z, Boss WF. Association of phosphatidylinositol kinase, phosphatidylinositol monophosphate kinase, and diacylglycerol kinase with the cytoskeleton and F-actin fractions of carrot (Daucus carota L.) cells grown in suspension culture. Plant Physiol 1992; 100:2116-2120.
6. Yamamoto K, Pardee JD, Reidler J et al. Mechanism of interaction of Dictyostelium severin with actin filaments. J Cell Biol 1982; 95:711-719.
7. Witke W, Hofmann A, Koppel B et al. The Ca2+-binding domains in nonmuscle type alpha-actinin: Biochemical and genetic analysis. J Cell Biol 1993; 121:599-606.
8. Fechheimer M, Taylor DL. Isolation and characterization of a 30,000-dalton calcium-sensitive actin cross-linking protein from Dictyostelium discoideum. J Biol Chem 1984; 259:4514-4520.
9. Doring V, Veretout F, Albrecht R et al. The in vivo role of annexin VII (synexin) characterization of an annexin VII deficient Dictyostelium mutant indicates an involvement in calcium-regulated processes. J Cell Sci 1995; 108:2065-2076.
10. Zhu Q, Liu T, Clarke M. Calmodulin and the contractile vacuole complex in mitotic cells of Dictyostelium discoideum. J Cell Sci 1993; 104:1119-1127.

11. Dharamsi A, Tessarolo D, Coukell B et al. CBP1 associates with the Dictyostelium cytoskeleton and is important for normal cell aggregation under certain developmental conditions. Exp Cell Res 2000; 258:298-309.

12. Huttenlocher A, Palecek SP, Lu Q et al. Regulation of cell migration by the calcium-dependent protease calpain. J Biol Chem 1997; 272:32719-32722.

13. Rabinovitch M. Professional and nonprofessional phagocytes: An introduction. Trends Cell Biol 1995; 5:85-87.

14. Kwiatkowska K, Sobota A. Signaling pathways in phagocytosis. BioEssays 1999; 21:422-431.

15. Indik ZK, Park JG, Hunter S et al. The molecular dissection of Fcγ receptor mediated phagocytosis. Blood 1995; 86:4389-4399.

16. Hampton MB, Kettle AJ, Winterbourn CC. Inside the neutrophil phagosome: Oxidants, myeloperoxidase, and bacterial killing. Blood 1998; 92:3007-3017.

17. Malik ZA, Denning GM, Kusner DJ. Inhibition of Ca^{2+} signaling by Mycobacterium tuberculosis is associated with reduced phagosome-lysosome fusion and increased survival within human macrophages. J Exp Med 2000; 191:287-302.

18. Ofek I, Goldhar J, Keisari Y et al. Nonopsonic phagocytosis of microorganisms. Annu Rev Microbiol 1995; 49:239-276.

19. Schutt C. Fighting infection: The role of lipopolysaccharide binding proteins CD14 and LBP. Pathobiology 1999; 67:227-229.

20. In: Roitt IM, ed. The basis of immunology. Essential Immunology. Oxford: Blackwell Scientific, 1994.

21. van Egmond MH, Hanneke van Vuuren AJ, van de Winkel JG. The human Fc receptor for IgA (FcαRI, CD89) on transgenic peritoneal macrophages triggers phagocytosis and tumor cell lysis. Immunol Lett 1999; 68:83-87.

22. Yokota A, Yukawa K, Yamamoto A et al. Two forms of the low affinity Fc receptor for IgE differentially mediate endocytosis and phagocytosis: Identification of the critical cytoplasmic domains. Proc Nat Acad Sci USA 1992; 89:5030-5034.

23. Sanchez-Mejorada G, Rosales C. Signal transduction by immunoglobulin Fc receptors. J Leuk Biol 1998; 63:521-533.

24. Indik ZK, Hunter S, Huang MM et al. The high affinity Fc gamma receptor (CD64) induces phagocytosis in the absence of its cytoplasmic domain: The gamma subunit of Fc gamma RIIIA imparts phagocytic function to Fc gamma RI. Exp Hematol 1994; 22:599-606.

25. Tuijnman WB, Capel PJ, van de Winkel JG. Human low affinity IgG receptor Fc gamma RIIa (CD32) introduced into mouse fibroblasts mediates phagocytosis of sensitized erythrocytes. Blood 1992; 79:1651-1656.

26. Park JG, Isaacs RE, Chien P et al. In the absence of other Fc receptors, Fc gamma RIIIA transmits a phagocytic signal that requires the cytoplasmic domain of its gamma subunit. J Clin Invest 1993; 92:1967-1973.

27. Indik ZK, Pan XQ, Huang MM et al. Insertion of cytoplasmic tyrosine sequences into the nonphagocytic receptor Fc gamma RIIB establishes phagocytic function. Blood 1994; 83:2072-2080.

28. Kimberly RP, Ahlstrom JW, Click ME et al. The glycosyl phosphatidylinositol-linked Fc gamma RIIIPMN mediates transmembrane signaling events distinct from Fc gamma RII. J Exp Med 1990; 171:1239-1255.

29. Salmon JE, Brogle EJ, Edberg JC et al. Fcgamma receptor III induces actin polymerization in human neutrophils and primes phagocytosis mediated by Fcgamma receptor II. J Immunol 1991; 146:997-1004.

30. Chuang FY, Sassaroli M, Unkeless JC. Convergence of Fc-gamma receptor IIA and Fc gamma receptor IIIB signaling pathways in human neutrophils. J Immunol 2000; 164:350-360.

31. Garcia-Garcia E, Rosales C. Signal transduction during Fc receptor-mediated phagocytosis. J Leuk Biol 2002; 72:1092-1108.

32. Smith LC, Azumi K, Nonaka M. Complement systems in invertebrates. The ancient alternative and lectin pathways. Immunopharmacology 1999; 42:107-120.

33. Ravetch JV, Clynes RA. Divergent roles for Fc receptors and complement in vivo. Annu Rev Immunol 1998; 16:421-432.

34. Brown EJ. Complement receptors and phagocytosis. Curr Opin Immunol 1991; 3:76-82.
35. Diamond MS, Garcia-Aguilar J, Bickford J et al. The I domain is a major recognition site on the leukocyte integrin Mac-1 (CD11b/CD18) for four distinct adhesion ligands. J Cell Biol 1993; 120:1031-1043.
36. Brown EJ. The role of extracellular matrix proteins in the control of phagocytosis. J Leuk Biol 1986; 39:579-591.
37. Pommier CG, Inada S, Fries LF et al. Plasma fibronectin enhances phagocytosis of opsonized particles by human peripheral blood monocytes. J Exp Med 1983; 157:1844-1854.
38. Oxvig C, Lu C, Springer TA. Conformational changes in tertiary structure near the ligand binding site of an integrin I domain. Proc Nat Acad Sci USA 1999; 96:2215-2220.
39. Chatila TA, Geha RS, Arnaout MA. Constitutive and stimulus-induced phosphorylation of CD11/CD18 leukocyte adhesion molecules. J Cell Biol 1989; 109:3435-3444.
40. Detmers PA, Wright SD, Olsen E et al. Aggregation of complement receptors on human neutrophils in the absence of ligand. J Cell Biol 1987; 105:1137-1145.
41. Allen LA, Aderem A. Mechanisms of phagocytosis. Curr Opin Immunol 1996; 8:36-40.
42. Blystone SD, Graham IL, Lindberg FP et al. Integrin alpha v beta 3 differentially regulates adhesive and phagocytic functions of the fibronectin receptor alpha 5 beta 1. J Cell Biol 1994; 127:1129-1137.
43. Stahl PD, Ezekowitz RA. The mannose receptor is a pattern recognition receptor involved in host defense. Curr Opin Immunol 1998; 10:50-55.
44. Devitt A, Moffatt OD, Raykundalia C et al. Human CD14 mediates recognition and phagocytosis of apoptotic cells. Nature 1998; 392:505-509.
45. Platt N, da Silva RP, Gordon S. Recognizing death: The phagocytosis of apoptotic cells. Trends Cell Biol 1998; 8:365-372.
46. Parnaik R, Raff MC, Scholes J. Differences between the clearance of apoptotic cells by professional and nonprofessional phagocytes. Curr Biol 2000; 10:857-860.
47. Meagher LC, Savill JS, Baker A et al. Phagocytosis of apoptotic neutrophils does not induce macrophage release of thromboxane B2. J Leuk Biol 1992; 52:269-273.
48. Fadok VA, Bratton DL, Konowal A et al. Macrophages that have ingested apoptotic cells in vitro inhibit proinflammatory cytokine production through autocrine/paracrine mechanisms involving TGF-beta, PGE2, and PAF. J Clin Invest 1998; 101:890-898.
49. Voll RE, Herrmann M, Roth EA et al. Immunosuppressive effects of apoptotic cells. Nature 1997; 390:350-351.
50. Savill J. Apoptosis. Phagocytic docking without shocking. Nature 1998; 392:442-443.
51. Fadok VA, Bratton DL, Rose DM et al. A receptor for phosphatidylserine-specific clearance of apoptotic cells. Nature 2000; 405:85-90.
52. Chung S, Gumienny TL, Hengartner MO et al. A common set of engulfment genes mediates removal of both apoptotic and necrotic cell corpses in C. elegans. Nature Cell Biol 2000; 2:931-937.
53. Roubey RA, Ross GD, Merrill JT et al. Staurosporine inhibits neutrophil phagocytosis but not iC3b binding mediated by CR3 (CD11b/CD18). J Immunol 1991; 146:3557-3562.
54. Zhou Z, Hartwleg E, Horvitz HR. CED-1 is a transmembrane receptor that mediates cell corpse engulfment in C. elegans. Cell 2001; 104:43-56.
55. Kaplan G. Differences in the mode of phagocytosis with Fc and C3 receptors in macrophages. Scand J Immunol 1977; 6:797-807.
56. Berridge MJ, Lipp P, Bootman MD. The versatility and universality of calcium signalling. Nat Rev Mol Cell Biol 2000; 1:11-21.
57. Bootman MD, Berridge MJ, Roderick HL. Calcium signalling: More messengers, more channels, more complexity. Curr Biol 2002; 12:R563-R565.
58. Bootman MD, Lipp P, Berridge MJ. The organisation and functions of local Ca^{2+} signals. J Cell Sci 2001; 114:2213-2222.
59. Pryor PR, Mullock BM, Bright NA et al. The role of intraorganellar Ca^{2+} in late endosome-lysosome heterotypic fusion and in the reformation of lysosomes from hybrid organelles. J Cell Biol 2000; 149:1053-1062.

60. Mao C, Kim SH, Almenoff JS et al. Molecular cloning and characterization of SCaMPER, a sphingoid Ca^{2+} release-mediating protein from endoplasmic reticulum. Proc Natl Acad Sci USA 1996; 93:1993-1996.
61. Schnurbus R, De Pietri Tnelli D, Grohovas F et al. Reevaluation of primary structure, topology and localization of SCaMPER, a putative intracellular Ca^{2+} channel activated by sphingosylphosphocholine. Biochem J 2002; 362:183-189.
62. Berridge MD. Capacitative calcium entry. Biochem J 1995; 312:1-11.
63. Cancela JM, van Coppenolle F, Galione A et al. Transformation of local Ca^{2+} spikes to global Ca^{2+} transients: The combinatioral roles of multiple Ca2+ releasing messengers. EMBO J 2002; 21:909-919.
64. Mingen O, Thompson JL, Shuttleworth TJ. Reciprocal regulation of capacitative and arachidonate-regulated noncapacitative Ca^{2+} entry pathways. J Biol Chem 2001; 276:35676-35683.
65. Montell C. Physiology, phylogeny and functions of the TRP superfamily of cation channels. Science's STKE 2001, http://stke.sciencemag.org/cgi/content/full/OC_sigtrans, 2001/90/re1.
66. Yue LX, Peng JB, Hediger MA et al. CaT1 manifests the pore properties of the calcium-release-activated calcium channel. Nature 2001; 410:705-709.
67. Voets T, Prenen J, Fleig A et al. CaT1 and the calcium release-activated calcium channel manifest distinct pore properties. J Biol Chem 2001; 276:47767-47770.
68. Tapper H. The secretion of preformed granules by macrophages and neutrophils. J Leukocyte Biol 1996; 59:613–622.
69. Brumell JH, Volchuk A, Sengelov H et al. Subcellular distribution of docking/fusion proteins in neutrophils, secretory cells with multiple exocytic compartments. J Immunol 1995; 155:5750–5759.
70. Floto RA, Mahaut-Smith MP, Somasundaram B et al. IgG-induced Ca^{2+} oscillations in differentiated U937 cells; a study using laser scanning confocal microscopy and coloaded Fluo-3 and Fura-red fluorescent probes. Cell Calcium 1995; 18:377-389.
71. Mandeville JT, Maxfield FR. Calcium and signal transduction in granulocytes. Curr Opin Hematol 1996; 3:63–70.
72. Pryor PR, Mullock BM, Bright NA et al. The role of intraorganellar Ca^{2+} in late endosome-lysosome heterotypic fusion and in the reformation of lysosomes from hybrid organelles. J Cell Biol 2000; 149:1053–1062.
73. Peters C, Mayer A. Ca^{2+}/calmodulin signals the completion of docking and triggers a late step of vacuole fusion. Nature 1998; 396:575–580.
74. Lundqvist-Gustafsson H, Gustafsson M, Dahlgren C. Dynamic Ca^{2+} changes in neutrophil phagosomes. A source for intracellular Ca^{2+} during phagolysosome formation?. Cell Calcium 2000; 27:353–362.
75. Berridge MJ. Inositol trisphosphate and calcium signalling. Nature 1993; 361:315-325.
76. Rosales C, Brown EJ. Signal transduction by neutrophil immunoglobulin G Fc receptors: Dissociation of intracytoplasmic calcium concentration rise from inositol 1,4,5,-trisphosphate. J Biol Chem 1992; 267:5265-5271.
77. Choi OH, Kim JH, Kinet J-P. Calcium mobilization via sphingosine kinase in signaling by the FcεRI antigen receptor. Nature 1996; 380:634-636.
78. Melendez AJ, Floto RA, Cameron AJ et al. A molecular switch changes the signalling pathway used by FcγRI antibody receptor to mobilise calcium. Curr Biol 1998; 8:210-221.
79. Melendez A, Floto RA, Gillooly DJ et al. FcγRI coupling to phospholipase D initiates sphingosine kinase-mediated calcium mobilization and vesicular trafficking. J Biol Chem 1998; 273:9393-9402.
80. Melendez AJ, Bruetschy L, Floto RA et al. Functional coupling of FcγRI to nicotinamide dinucleotide phosphate (reduced form) oxidative burst and immune complex trafficking requires the activation of phospholipase D1. Blood 2001; 98:3421-3428.
81. Melendez AJ, Khaw AK. Dichotomy of Ca^{2+} signals triggered by different phospholipid pathways in antigen stimulation of human mast cells. J Biol Chem 2002; 277:17255-17262.
82. Rosales C, Jones SL, McCourt D et al. Bromophenacyl bromide binding to the actin-bundling protein l-plastin inhibits inositol trisphosphate-independent increase in Ca^{2+} in human neutrophils. Proc Natl Acad Sci USA 1994; 91:3534-3538.
83. Olivera A, Rosenthal J, Spiegel S. Effect of acidic phospholipids on sphingosine kinase. J Cell Biochem 1996; 60:529-537.

84. Young JD, Ko SS, Cohn ZA. The increase in intracellular free calcium associated with IgG gamma 2b/gamma 1 Fc receptor-ligand interactions: Role in phagocytosis. Proc Nat Acad Sci USA 1984; 81:5430-5434.

85. Kobayashi K, Takahashi K, Nagasawa S. The role of tyrosine phosphorylation and Ca^{2+} accumulation in Fc gamma-receptormediated phagocytosis of human neutrophils. J Biochem (Tokyo) 1995; 117:1156-1161.

86. Della Bianca V, Grzeskowiak M, Rossi F. Studies on molecular regulation of phagocytosis and activation of the NADPH oxidase in neutrophils. IgG- and C3bmediated ingestion and associated respiratory burst independent of phospholipid turnover and Ca^{2+} transients. J Immunol 1990; 144:1411-1417.

87. Lew DP, Andersson T, Hed J et al. Ca^{2+}-dependent and Ca^{2+}-independent phagocytosis in human neutrophils. Nature 1985; 315:509-511.

88. Bengtsson T, Jaconi ME, Gustafson M et al. Actin dynamics in human neutrophils during adhesion and phagocytosis is controlled by changes in intracellular free calcium. Eur J Cell Biol 1993; 62:49-58.

89. Jaconi ME, Lew DP, Carpentier JL et al. Cytosolic free calcium elevation mediates the phagosome lysosome fusion during phagocytosis in human neutrophils. J Cell Biol 1990; 110:1555-1564.

90. Greenberg S, el Khoury J, di Virgilio F et al. Ca^{2+}-independent F-actin assembly and disassembly during Fc receptor-mediated phagocytosis in mouse macrophages. J Cell Biol 1991; 113:757-767.

91. Zimmerli S, Majeed M, Gustavsson M et al. Phagosome-lysosome fusion is a calcium-independent event in macrophages. J Cell Biol 1996; 132:49-61.

92. Edberg JC, Lin CT, Lau D et al. The Ca^{2+} dependence of human Fc gamma receptor-initiated phagocytosis. J Biol Chem 1995; 270:22301-22307.

93. Stendahl O, Krause KH, Krischer J et al. Redistribution of intracellular Ca^{2+} stores during phagocytosis in human neutrophils. Science 1994; 265:1439–1441.

94. Sawer DW, Sullivan JA, Mandell GL. Intracellular free calcium localization in neutrophils during phagocytosis. Science 1985; 230:663-666.

95. Bengtsson T, Jaconi ME, Gustafson M et al. Actin dynamics in human neutrophils during adhesion and phagocytosis is controlled by changes in intracellular free calcium. Eur J Cell Biol 1993; 62:49–58.

96. Harris HE, Weeds AG. Plasma gelsolin caps and severs actin filaments. FEBS Lett 1984; 177:184-188.

97. Yin HL, Albrecht JH, Fattoum A. Identification of gelsolin, a Ca^{2+}-dependent regulatory protein of actin gel-sol transformation, and its intracellular distribution in a variety of cells and tissues. J Cell Biol 1981; 91:901-906.

98. Serrander L, Skarman P, Rasmussen B et al. Selective inhibition of IgG-mediated phagocytosis in gelsolin-deficient murine neutrophils. J Immunol 2000; 165:2451-2457.

99. Allen LA, Aderem A. Molecular definition of distinct cytoskeletal structures involved in complement- and Fc receptor-mediated phagocytosis in macrophages. J Exp Med 1996; 184:627-637.

100. Allen LH, Aderem A. A role for MARCKS, the alpha isozyme of protein kinase C and myosin I in zymosan phagocytosis by macrophages. J Exp Med 1995; 182:829-840.

101. Larsen EC, DiGennaro JA, Saito N et al. Differential requirement for classic and novel PKC isoforms in respiratory burst and phagocytosis in RAW 264.7 cells. J Immunol 2000; 165:2809-2817.

102. Greenberg S, Chang P, Silverstein SC. Tyrosine phosphorylation is required for Fc receptor-mediated phagocytosis in mouse macrophages. J Exp Med 1993; 177:529-534.

103. Zheleznyak A, Brown EJ. Immunoglobulin-mediated phagocytosis by human monocytes requires protein kinase C activation. Evidence for protein kinase C translocation to phagosomes. J Biol Chem 1992; 267:12042-12048.

104. Dekker LV, Leitges M, Altschuler G et al. Protein kinase C-beta contributes to NADPH oxidase activation in neutrophils. Biochem J 2000; 347:285-289.

105. Melendez AJ, Harnett MM, Allen JM. Differentiation dependent switch in protein kinase C isoenzyme activation by FcgammaRI, the human high-affinity receptor for immunoglobulin G. Immunology 1999; 96:457-464.

106. Brumell JH, Howard JC, Craig K et al. Expression of the protein kinase C substrate pleckstrin in macrophages: Association with phagosomal membranes. J Immunol 1999; 163:3388-3395.

107. Li J, Aderem A. MacMARCKS, a novel member of the MARCKS family of protein kinase C substrates. Cell 1992; 70:791-801.
108. Hartwig JH, Thelen M, Rosen A et al. MARCKS is an actin filament crosslinking protein regulated by protein kinase C and calciumcalmodulin. Nature 1992; 356:618-622.
109. Carballo E, Pitterle DM, Stumpo DJ et al. Phagocytic and macropinocytic activity in MARCKS-deficient macrophages and fibroblasts. Am J Physiol 1999; 277:C163-173.
110. Underhill DM, Chen J, Allen LA et al. MacMARCKS is not essential for phagocytosis in macrophages. J Biol Chem 1998; 273:33619-33623.
111. Zhu Z, Bao Z, Li J. MacMARCKS mutation blocks macrophage phagocytosis of zymosan. J Biol Chem 1995; 270:17652-17655.
112. Colombo MI, Beron W, Stahl PD. Calmodulin regulates endosome fusion. J Biol Chem 1997; 272:7707-7712.
113. Wickner W, Haas A. Yeast homotypic vacuole fusion: A window on organelle trafficking mechanisms. Annu Rev Biochem 2000; 69:247-275.
114. Zimmerli S, Majeed M, Gustavsson M et al. Phagosome-lysosome fusion is a calcium-independent event in macrophages. J Cell Biol 1996; 132:49-61.
115. Jaconi ME, Lew DP, Carpentier JL et al. Cytosolic free calcium elevation mediates the phagosome-lysosome fusion during phagocytosis in human neutrophils. J Cell Biol 1990; 110:1555-1564.
116. Downey GP, Botelho RJ, Butler JR et al. Phagosomal maturation, acidification, and inhibition of bacterial growth in nonphagocytic cells transfected with FcγRIIA receptors. J Biol Chem 1999; 274:28436-28444.
117. Wilsson A, Lundqvist H, Gustafsson M et al. Killing of phagocytosed Staphylococcus aureus by human neutrophils requires intracellular free calcium. J Leuk Biol 1996; 59:902-907.
118. Jahraus A, Egeberg M, Hinner B et al. ATP-dependent membrane assembly of F-actin facilitates membrane fusion. Mol Biol Cell 2001; 12:155-170.
119. Ferrari G, Langen H, Naito M et al. A coat protein on phagosomes involved in the intracellular survival of mycobacteria. Cell 1999; 97:435-447.
120. Greenberg S, el Khoury J, di Virgilio F et al. Ca^{2+}-independent F-actin assembly and disassembly during Fc receptor-mediated phagocytosis in mouse macrophages. J Cell Biol 1991; 113:757-767.
121. Ernst JD. Annexin III translocates to the periphagosomal region when neutrophils ingest opsonized yeast. J Immunol 1991; 146:3110-3114.
122. Majeed M, Perskvist N, Ernst JD et al. Roles of calcium and annexins in phagocytosis and elimination of an attenuated strain of Mycobacterium tuberculosis in human neutrophils. Microb Pathog 1998; 24:309-320.
123. Malik ZA, Iyer SS, Kusner DJ. Mycobacterium tuberculosis phagosomes exhibit altered calmodulin-dependent signal transduction: Contribution to inhibition of phagosome-lysosome fusion and intracellular survival in human macrophages. J Immunol 2001; 166:3392-3401.

CHAPTER 10

Phagosome Maturation

William S. Trimble and Marc G. Coppolino

Abstract

P hagocytosis is the process used by eukaryotic cells to engulf and ingest foreign particles. In lower eukaryotes this process is mainly used for food uptake while in multicellular organisms phagocytosis is the primary mechanism used to fight infection. Phagocytosis of microbes typically leads to their killing as the organelle of ingestion, the phagosome, acquires anti-microbial properties through a process termed "phagosomal maturation". Since many of these microbicidal properties are also typical of the lysosome, the mature phagosome has been called a phagolysosome and parallels have been drawn between the maturation process and the endocytic pathway. Importantly, some microbes have evolved mechanisms to abrogate the maturation process, allowing them to persist as intracellular parasites within the very cells charged with the task of eliminating them. In this chapter we will focus on the molecular mechanisms involved in the maturation of the mammalian macrophage phagosome, especially those formed by activation of the Fc receptor, and how some intracellullar parasites derail this maturation process.

Endocytosis: Snares, Rabs and Tethers

The Endocytic Pathway

Phagocytosis and phagocytic maturation have been likened to the process of endocytosis, with the phagosome being internalized like an endocytic vesicle and gradually maturing into a lysosome by sequential delivery of components of the endocytic pathway (Fig. 1). The endocytic pathway is a dynamic assembly of membranous compartments, each bearing unique biochemical and physiological properties (for review see ref. 1). Internalization of membrane proteins, receptors and their ligands from the plasma membrane can occur by clathrin-dependent and clathrin-independent mechanisms and these are delivered to early endosomes. In early endosomes, proteins are sorted for degradation towards late endosomes and lysosomes, or are recycled to the cell surface after trafficking through perinuclear recycling endosomes. Ubiquitination of membrane proteins can contribute to their internalization and lead to their degradation in the multivesicular body. Movement of membranes and membrane proteins through the endocytic pathway occurs via the formation of vesicles and tubules that bud off individual compartments and fuse with the next. Correct targeting of these transport compartments likely involves the coordinated action of the SNARE and rab proteins.

Molecular Mechanisms of Phagocytosis, edited by Carlos Rosales. ©2005 Eurekah.com and Springer Science+Business Media.

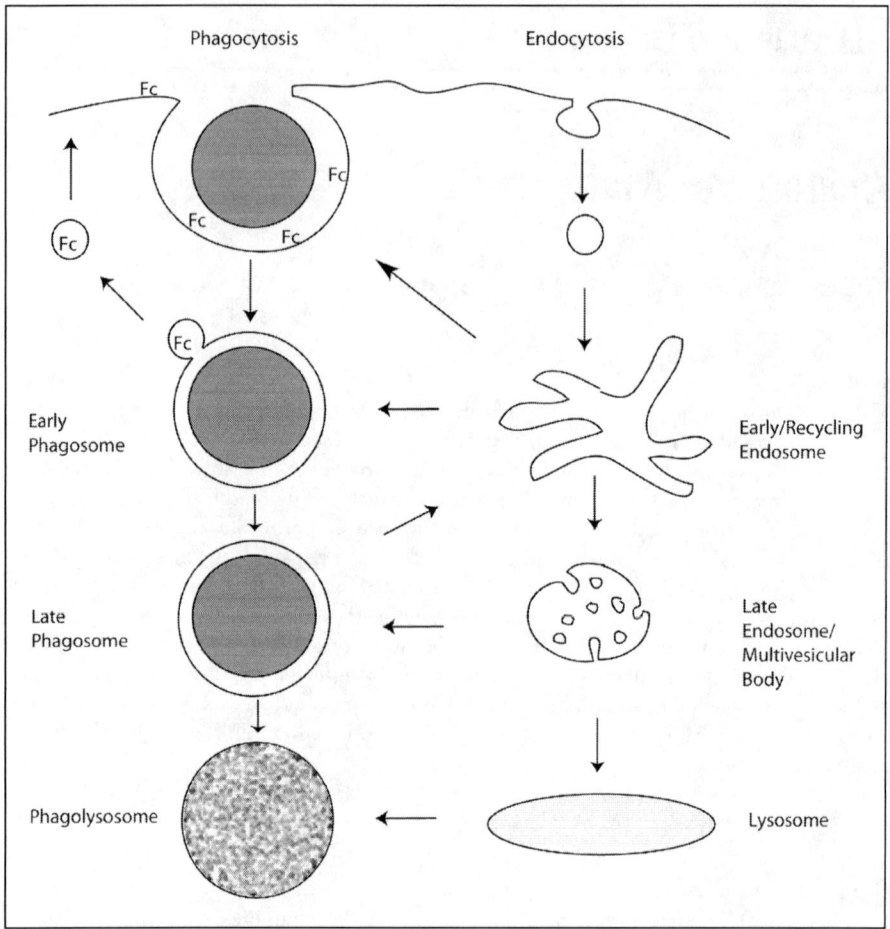

Figure 1. Phagosome and endosome maturation. Phagosome maturation (left) parallels that of endosome maturation (right) due to the delivery of endosomal and lysosomal membranes and contents to the maturing phagosome. This may occur through vesicular transport intermediates or direct fusion of organelles (complete fusion or 'kiss and run') with compensatory removal of membranes and proteins simultaneously occurring.

SNARE Proteins

The mechanisms controlling the fusion and fission of transport vesicles in the endosomal system are now becoming understood, in large part due to extrapolation from earlier studies involving the study of synaptic vesicle fusion in the nervous system. Most current models of membrane fusion have evolved from the SNARE hypothesis, first proposed by J. Rothman[2] in which the synaptic vesicle protein VAMP[3] (called a vesicle (v)-SNARE) and the proteins syntaxin[4] and SNAP25[5] (called t-SNAREs due to their predominance on target membranes) bind to each other to drive the fusion of the membranes. After the fusion event has occurred the SNARE proteins remain in a stable complex[6] until an ATPase known as NSF[7] and its adaptor protein, called αSNAP,[8] bind to the complex and uncouple it to allow additional rounds of fusion. In vitro studies have shown that the interaction of SNARE proteins alone is necessary and sufficient for fusion.[9]

Many isoforms of SNARE proteins have been identified and most can be categorized as distant members of the VAMP, syntaxin or SNAP-25 subfamilies. All SNARE proteins contain one or two helical domains of about 60 amino acids with the propensity form coiled coils and these have been called SNARE domains.[10] In mammals, at least 18 syntaxin-like tSNAREs have been reported to date. Similarly large numbers of VAMP-like vSNAREs have been identified while the SNAP-25 family appears to be much smaller, although some of the syntaxin-like proteins appear to represent the N- and C-terminal halves of SNAP-25 and may contribute its function for specific trafficking steps. Individual isoforms of the v- and t-SNARE proteins are thought to reside on specific membranes and, even though their binding in vitro appears promiscuous,[11] their correct pairing may contribute to proper vectoral traffic of membranes within cells.[12]

In the endocytic pathway, the early endosomes possess syntaxins 7, 8 and 13 and VAMP 3 and 8. Late endosomes possess syntaxins 7 and 8 and VAMP 7 and 8 while syntaxin 7 and 8 are also found on lysosomes.[13] The role of each of these proteins in traffic to and from each compartment has not been fully elucidated, but given the dynamic nature of the endocytic organelles it is likely that SNAREs would be in dynamic equilibrium between the organelle of their destination and that of their source. VAMP3 likely serves as the vesicle SNARE on endocytic vesicles recycling from endosomes to the plasma membrane[14] where it engages syntaxin 4 and SNAP23 to deliver new membranes from the endosome to the cell surface.[15] VAMP3 does predominate in both the early and recycling endosome compartments and it is likely retrieved from the surface and may therefore participate, in conjunction with syntaxin 13, in the fusion of endocytic vesicles with early endosomes. Similarly, syntaxin 7 is found predominantly in late endosomes where it appears to function as the tSNARE for incoming transport vesicles from the early endosome,[16] but it is also found distributed in the early endosomes and lysosomes.

Rab Proteins

Rab proteins are members of a family of small GTPases present in all eukaryotic species. They are membrane associated proteins due to an isoprenoid moiety at their C-termini and are found on the cytoplasmic face of transport vesicles and organelles. There are at least 60 different isoforms of rab proteins, each thought to regulate a specific step within the cell, and some of these appear to be tissue specific as well.[17] In general terms the rabs are thought to act as molecular switches that cycle between the active GTP-bound state and the inactive GDP-bound state. Their activity is controlled by two other classes of proteins: guanine nucleotide exchange factors which turn on the rabs by promoting the exchange of GDP for GTP; and GTPase activating proteins, effectors that turn off the rabs by enhancing the otherwise low intrinsic GTPase activity of the rabs to promote hydrolysis of GTP to GDP. In the endocytic pathway, the two major rab forms involved in internalization of proteins are rab5 and rab7 while rabs 4 and 11 appear to function in recycling of proteins back to the cell membrane. Rab5 is involved in the traffic of endocytic vesicles to early endosomes and in endosome-endosome fusion. Rab7 controls traffic from early to late endosomes and to lysosomes. One rab5 effector is early endosomal antigen 1 (EEA1), a 162 kDa membrane bound protein component specific to early endosomes that is essential for fusion between early endocytic vesicles.[18] EEA1 contains zinc-finger-like domains, similar to those found in nucleic acid binding proteins, located in its amino and carboxyl-terminal domains. The carboxyl-terminal zinc-finger-like-domain is conserved in several other nonnuclear proteins and is an authentic zinc-binding domain capable of direct binding to phosphatidylinositol-3-phosphate (PI3P), the product of phosphatidyl 3-kinase (PI3K). Moreover, EEA1 localization to endosomes is dependent on PI3K activity, and inhibitors of this enzyme cause EEA1 to dissociate from early endosomes. Rab5 is present on endocytic vesicles and is also abundant on early endosomes. EEA1 bridges these two compartments by two spatially separate rab5 binding domains, possibly providing the tethering function necessary to allow SNARE interaction.

Formation of a Phagosome

Phagocytic Receptor Clustering Activates Signaling Pathways That Lead to Phagocytosis

Phagocytosis is intiated by the engagement and clustering of phagocytic receptors. Many receptors are capable of inducing phagocytosis, including the Toll-like pattern recognition receptors detecting common bacterial surface components such as lipopolysaccharides and lipoproteins. In contrast to these so-called "nonopsinic" receptors, other receptors recognize foreign particles in the context of immunoglobulin or complement fragment C3bi immune molecules. These opsinins bind to the surface of particles allowing the engagement of receptors recognizing the Fc portion of antibodies (Fc receptors) and complement (CR receptors) respectively. The structure and functional aspects of these and other receptors have been reviewed elsewhere in this book and will not be dealt with here. For the rest of this chapter we will focus on Fc-mediated phagocytosis since this has been most extensively characterized. Activation of Fc receptors leads to the tyrosine phosphorylation of the receptors, or associated subunits, by members of the src kinase family. Downstream of src phosphorylation are a number of biochemical events involving adaptor proteins (such as Slp-76, LAT, SLAP), kinases such as Syk and Nck, lipid kinases and GTPases. Detailed discussions of the signaling pathways leading to phagosome formation and GTPase function in phagocytosis are presented elsewhere in this book.

Actin Assembly at the Base of the Nascent Phagosome

A predominant feature of phagocytosis is the accumulation of actin and associated proteins at the base of the forming phagosome. It has long been recognized that remodeling of actin filaments is required for efficient particle internalization and studies of the regulation of actin reorganization have identified several proteins and lipids with direct roles in this process. One of the earliest detectable biochemical events downstream of Fc receptor phosphorylation is activation of Syk kinase. Syk can stimulate actin remodeling and its activity is required for particle internalization.[19-23] The signaling pathway that links the activation of Syk to actin remodeling during phagocytosis appears to involve the formation of a molecular complex containing several proteins that are associated with the regulation of actin reorganization, namely Fyb/src-like adaptor protein (SLAP), SLP-76, Nck, vasodilator-stimulated phosphoprotein (VASP) and Wiskott-Aldrich syndrome protein (WASP).[24] This complex forms early during the engulfment of particles and may function to synchronize the localization of key mediators of actin remodeling, for example profilin and Arp2/3, at sites of phagosome formation.

The seven subunit Arp2/3 complex has a primary role in the nucleation of actin filaments in many cellular functions.[25,26] The Arp2/3 complex has been shown to localize to both FcγR- and CR3-mediated phagosomes and is necessary for particle ingestion via both of these receptors.[27] This indicates that nucleation of actin structures is required for phagosome formation and a candidate regulator of Arp2/3-directed actin nucleation is WASP, which can bind and activate Arp2/3.[26,28] WASP was found to be recruited to the forming FcγR-mediated phagosomes in macrophages.[29] Interestingly, in WAS patients the levels of WASP are reduced and macrophages from these patients display impaired FcγR-dependent phagocytosis.[29] It is not known if reduced WASP levels act primarily to lower the level of Arp2/3 activation in these cells, but it has been shown in other systems that Cdc42 and phosphatidylinositol-4,5-bisphosphate (PIP$_2$) can activate N-WASP and lead to activation of Arp2/3.[30] It is thus possible that WASP function links these pathways to the nucleation of actin filaments during phagocytosis.

Actin-severing proteins and proteins that depolymerize actin filaments have also been implicated in phagocytosis. For example, the actin-severing and capping protein gelsolin has been shown to localize to phagosomes. Neutrophils from gelsolin[-/-] mice have defective FcγR-mediated phagocytosis; however, phagosome-associated actin reorganization is apparently normal in these cells[31] and it is not currently understood how a lack of gelsolin might alter phagocytic capacity independently of actin remodeling. There is also some evidence suggesting that the actin-depolymerizing activity of cofilin is required for phagocytosis. This includes the facts that cofilin can associate with phagosomes in *Dictyostelium*[32] and that microinjection of antibodies to cofilin impedes phagocytosis in neutrophils.[33]

The strength of actin filaments can be enhanced by bundling or cross-linking of the filaments. This activity is carried out by several proteins, some of which have been implicated in phagocytosis. The actin-bundling protein α-actinin and the actin-cross-linking protein MARCKS (myristoylated alanine-rich C kinase substrate) have been found to associate with forming phagosomes, but the physiological requirement for these proteins during phagocytosis has not been established.[34-37] In contrast to α-actinin and MARCKS, a lack of the actin-bundling protein ABP-120 has been shown to cause a significant impairment in phagocytosis in *Dictyostelium*.[34,38,39]

The list of actin-regulating molecules that can be found on forming phagosomes also includes proteins that are known regulators of actin function in focal adhesions, for example talin, vinculin, paxillin and focal adhesion kinase (FAK).[34,40-42] In each of these cases, it is not clear what the function of the protein is during phagocytosis or if the involvement is physiologically relevant. Resolution of the physiological significance of these numerous actin regulating proteins and the phagocytic signaling pathways in which they participate stands to be gained from analyses of the function of these proteins in meaningful temporal and physiological contexts.

Phosphatidylinositol phosphate (PIP) lipids are critical regulators of actin structures and different PIP species are transiently associated with phagosomes as they form. Phosphatidylinositol-4-phosphate 5-kinase (PI5K) is recruited to nascent phagosomes immediately after engagement of receptors and likely leads to the production of PIP_2 from $PI4P$.[43,44] Indeed, PIP_2 production has been visualized in real time by the use of green fluorescent proteins fused to the lipid binding PH domains from a variety of proteins. Using the PH domain of PLCγ that has a high degree of specificity for PIP_2, Grinstein and colleagues showed that the accumulation of PIP_2 coincided with the recruitment of PI5K. PIP_2 is important in actin recruitment and polymerization, but as phagosome formation proceeds it is necessary to disassemble the actin mesh surrounding the particle to allow its internalization. A remarkable biphasic change was observed with the initial accumulation of PIP_2 followed by a loss of PIP_2 at the base of the forming phagosome and the accumulation of phospholipase C. Not surprisingly, it was found that the phospholipase C recruitment correlated with the accumulation of diacylglycerol, as detected by a fusion between GFP and the C1 domain of PKCδ, even before the sealing of the nascent phagosome. It seems likely that temporal coordination of PI phosphorylation and hydrolysis is involved in the initial recruitment of actin to the base and sides of the cup, and its subsequent release from these sites as the phagosome is internalized.

Substantial evidence now indicates that full pseudopod extension, particularly in the case of particles larger than 2-3 μm in diameter, and closure of the phagosome, occur in a class I PI3K-dependent manner.[45-47] During phagocytosis in the presence of inhibitors of PI3K, pseudopod extension may not be optimal, but phagosomal cup-associated actin polymerization is relatively normal.[45] These findings indicate that PI3K-regulated actin remodeling is not essential for phagocytosis and that PI3K may influence two distinct aspects of phagocytosis: actin polymerization and membrane remodeling. Both of these activities are ultimately required to support the formation of complete pseudopodia and efficient particle engulfment.

Expansion of the Membrane to Produce the Phagosome

Despite the accumulation of actin within the membrane ruffles that will engulf the particle, it is unlikely that actin assembly is the only driving force required to engulf the particle. This notion is supported by observations of pseudopodial extension that can occur in the absence of actin polymerization.[45] Many studies have now contributed to a model of phagocytosis that involves the delivery of intracellular membranes to the surface of the cell during particle internalization.[48-50] Evidence that membrane addition by the fusion of endomembranes to the cell surface is crucial for phagocytosis came from studies in which the SNARE proteins were perturbed. Impairment of membrane addition by the general inhibition of SNARE function using dominant-negative forms of NSF[51] abrogated the internalization of opsinized particles without affecting the binding to, or signaling from, Fc receptors. This suggested that expansion of the plasma membrane was necessary for the internalization of particles in phagosomes. More specific inhibition of the VAMP3 by the addition of Tetanus toxin, a VAMP-specific protease, also blocked internalization, supporting a role for this specific SNARE in membrane expansion.[50] Together, these studies indicated that membrane addition was required to expand the cell surface to compensate for phagocytosis, but left open the questions of where the new membranes come from and where they are added.

Partial delineation of the membrane traffic pathway involved in phagocytosis came with the demonstration that VAMP3 proteins function to deliver vesicles from the recycling endosome to the site of pseudopodial extension. Indeed, GFP-labeled VAMP3 was not only observed to be added to the cell surface during phagocytosis, but it was focally delivered to the nascent phagosome during the growth of the membranous pseudopods.[52] More recently it has been shown that ARF6 is responsible for signaling the delivery of the VAMP3-containing vesicles to the forming phagosome.[53] Intererestingly, mice bearing a deletion in the VAMP3 gene displayed only a small delay in phagocytosis of zymosan but otherwise phagocytosed normally.[54] The discrepancy between knockout and acute experiments may suggest that compensatory or redundant mechanisms are capable of overcoming the lack of VAMP3, but these await identification. Other recent studies have also implicated the endoplasmic reticulum as a source of endomembrane being added to the phagosome to permit its growth.[55] Despite the clear evidence that ER components do contribute to the phagosomal membrane in macrophages, there remains no evidence that the ER contribution is necessary for the initial formation of the phagosome. In addition, this organelle does not appear to contribute at all to the formation of phagosomes in neutrophils. ER membranes may, however, make important contributions to antigen presentation and to the survival of particular pathogens such as *Legionella pneumophila* (see below).

After the pseudopodia fully surround the foreign particle, these somehow fuse with each other, extracellularly, to seal the particle into the phagosome. No mechanisms have been proposed to explain how this event may occur although one possibility is that the actin filaments found in the protrusive pseudopodia may be used to provide a contractile "purse-string" function. Two motor proteins have been identified that could play this role in pseudopodia formation and sealing: myosin X and myosin II. Myosin X is widely expressed in many tissues in mammalian cells but is unique among the myosins in that the tail of myosin X contains three PH domains that may be phosphoinositide-binding regions. The cargo of myosin X is presently unknown, but it is known that this motor protein is responsible for some intrafilapodial motility.[56] Myosin X is concentrated at the tips of filopodia and myosin X rich regions travel forward and backward along filopods.[56] Recently myosin X has been shown to be a downstream effector of PI3K in phagocytosis, although its role in this process is not known.[57] Myosin X may be important in delivery of vesicle into the growing pseudopodia. Studies by Swanson and colleagues have shown that ML7, an inhibitor of myosin light chain kinase—and therefore an indirect inhibitor of myosin II, impaired

Table 1. Similar molecules are associated with early endosomes and phagosomes and late endosomes and phagosomes

Early Endosome/Early Phagosome	Late Endosome/Late Phagosome
EEA1	rab7
Rab5	rab9
PI(3)P	mannose-6-P Receptor
Syntaxin 13	syntaxin 7
VAMP3	LAMP °
Transferrin receptor	lysobisphosphatidic acid
Annexin 1	cathepsin D

phagocytosis.[58] This treatment resulted in phagocytic cups with actin-rich protrusions that surrounded IgG-opsinized red blood cells, but these did not squeeze the particles the way that untreated cups did, and the cups were more open and less circular. Whether myosin II is capable of a purse-string closure of the phagosome remains to be determined.

Phagosome Maturation

Once internalized, the phagosomal membrane undergoes a progressive maturation, and its composition changes, first containing components of the plasma membrane, then early endosomes and subsequently acquiring markers found in the late endosome and lysosome compartments.[59] In general, the earliest markers seen within the first few minutes after sealing of the phagosome include rab5, EEA1, PIP$_2$, components of the vacuolar ATPase and other components of the early endosomes. Consistent with the acquisition of the proton pump, acidification of the phagosome rapidly occurs. Loss of the plasma membrane and endosomal proteins then coincides with the acquisition of late endosome/lysosome components. These include lysosome-associated membrane proteins (LAMPs), rab7, cathepsin D and other hydrolytic enzymes. During this period the pH of the phagosome continues to drop to levels consistent with the lysosome (pH ≤ 5). (See Table 1). The timing of these steps appears to depend on the type of particle being ingested and the nature of the receptors being engaged. Latex beads are particularly slow and this slow time-course has been useful in characterizing the mechanisms of membrane delivery to the phagosome.

In vitro studies have shown that phagosome-endosome and phagosome-lysosome fusions are dependent on (NSF),[60,61] indicating a requirement for SNARE proteins throughout this process. Moreover, it has been observed that the efficiency of defined fusion events with the isolated phagosome changes over time, suggesting that temporal changes in its membrane composition influence its capacity for fusion. Early phagosomes isolated within 20 min of the onset of phagocytosis fuse readily with endosomes but poorly with lysosomes, while phagosomes that have matured for 2 hours in vivo fuse most efficiently with lysosomes but not at all with endosomes.[62,63]

One explanation for this temporal change in fusion capacity is that SNARE proteins such as syntaxin, which are localized to specific cellular compartments, are required at different steps in phagosomal maturation in vivo. Since the initial phagosome would have the SNARE composition of the plasma membrane, it would need to acquire syntaxin 13 to be capable of fusion with the late endosomes. Similarly, syntaxin 7 would be required for fusion with lysosomal vesicles. Indeed, we have found that syntaxin 13 and syntaxin 7 reside in different compartments in phagocytic cells and that they accumulate in the phagosome at different rates. Syntaxin 13 accumulated very early, within 5 min of the start of phagocytosis, and rapidly disappeared from the phagosome following particle internalization. In contrast, syntaxin 7 did not begin to appear until after 10 min and continued to accumulate during the course of maturation. The time course of syntaxin 13 accumulation is consistent with that previously observed for VAMP-3[52] and nascent phagosomes were often found to be positive for syntaxin 13-GFP before the apparent closure of the phagocytic cup. Interestingly, while inhibition of VAMP-3 by tetanus toxin impaired phagocytosis by limiting the extension of pseudopods, inhibition of syntaxin 13 function appeared to have no detectable effect on phagocytosis per se, but interfered with the subsequent steps required for maturation of the phagosome. This is consistent with the concept that the acquisition of syntaxin 13 at the forming phagosome is not needed for that step, but likely permits the phagosome to receive membranes from later stages in the endocytic pathway.

Each step in the endocytic pathway likely requires distinct sets of syntaxin proteins to receive the appropriate transport vesicles. For phagosome maturation to mimic the stepwise progression through the endocytic pathway would therefore require the removal of the existing syntaxins with replacement by the syntaxin responsible for the next step. Based on these observations we would hypothesize that syntaxin 13 serves to facilitate early endosome fusion events, while syntaxin 7 would be required to permit the fusion of vesicles from the late endosome/lysosome compartments. This would be consistent with in vitro studies showing that "young" phagosomes fuse efficiently with endosomes, while "old" phagosomes can fuse best with lysosomes.

The directed delivery of vesicle components to the phagosome also likely requires the action of rab proteins. Both in vitro and in vivo studies have provided evidence for a role of rab5 in early steps of phagosome maturation. In vitro, rab5 is required for the fusion of purified early endosomal membranes with purified newly formed phagosomes.[64] Maturation of the newly formed phagosome in vivo also correlates with the transient acquisition of rab5 as it is recruited from the cytoplasm.[65] Rab5 appears to control the duration of fusion events, with transient fusions called "kiss and run" between cellular organelles and the phagosome affecting the kinetics of the intermixing of their membrane contents[66] (discussed below). Interestingly, while both rab5 and PI3K are required for maturation of phagosomes, and PI3P production appears to be responsible for the recruitment of the rab5 binding protein EEA1, recruitment of rab5 itself appears to be independent of PI3K activation. Moreover, rab5 appears to act on both PI3K-dependent and PI3K-independent effectors, both of which are necessary to promote phagosome maturation.[65]

Rab7 appears to be important in late stages of endosomal traffic, although less is known about its roles in phagocytosis. Dominant-negative and constitutively active rab7 mutants have revealed that this protein controls transit from early to late endosomes[67,68] although it is also found on lysosomes. Overexpression of dominant-negative forms of rab7 in Dictyostelium prevent maturation of the phagosome[69] although the precise mechanisms are not known.

Although it is clear that membrane fusion is required for the delivery of new components to the phagosome, Desjardins and colleagues have proposed that complete fusion of the endocytic compartments to the phagosome may not occur. In a process coined "kiss and run", they suggest that the endocytic compartments fuse only transiently with the phagosome, allowing selective delivery of specific molecules into the phagosome.[70] The kiss and run model suggests that phagosomes and endosomes move along cytoskeletal elements and interact at focal points were they can fuse with each other. In support of this model, it has been shown that small soluble molecules can be delivered into phagosomes more quickly than larger molecules, arguing against full fusion of late endosomes or lysosomes with the phagosome. Similar mechanisms have been suggested for the release of neurotransmitters in the nerve terminal and this may permit a more efficient development of an end-stage organelle without the need for complete recycling of membrane components.

Regardless of whether membrane components and contents are delivered by full fusion of kiss-and-run, some of these proteins and membranes must be removed from the phagosome to allow it to maintain its size. Some phagosomal proteins will be recycled back to the prior compartment while others will be sorted to the lysosomal pathway for degradation. At the earliest stages of phagocytosis the phagosome contains membrane proteins such as the transferrin receptor, as well as the Fc receptor. The pinching of membranes from the surface of the phagosome likely involves the machinery responsible for vesicle formation at other stages of the endocytic pathway. Interestingly, the recycling of transferrin receptors and CD18 from maturing phagosomes was found to be independent of the activity of dynamin I.[71] It remains unclear whether clathrin or other coat proteins play a direct role in this process. FcγRIIA receptors are also ubiquitinated and while ubiquitination appears to be important for receptor endocytosis, mutations that prevent this do not impair phagosome maturation,[72] suggesting that this modification is not involved in removing receptors from the phagosome. COP-mediated coat formation has also been implicated in phagosome formation and maturation, although inhibition of COPI leads to only a slow and partial blockade of phagocytosis,[73] suggesting that this may not be directly involved in membrane recycling or that redundancy exists in these processes.

The Role of the Cytoskeleton during Phagosomal Maturation

Compartments of the endocytic/exocytic pathway occupy characteristic locations within cells. The intracellular location of these compartments, as well as the traffic between them, is dependent upon the integrity of the microtubule network. In an analogous manner, the maturing phagosomal compartment is dependent upon the microtubule network for its transformation into a phagolysosome. After a phagosome is formed at the cell periphery, it is transported toward the interior of the cell in a microtubule-dependent fashion.[74-76] While phagosome motility has been observed in both directions along microtubules, dynein-mediated minus-end-directed movement seems to predominate over kinesin-mediated plus-end-directed movement, consistent with the centripetal movement of phagosomes observed in cells.[75] The association of phagosomes with microtubules appears to be necessary for their maturation, as indicated by decreases in phagosomal acquisition of late-endosomal and lysosomal markers, including LAMP-2,[77,78] and by inhibition of phagosome-lysosome fusion[79] in cells treated with microtubule depolymerizing agents. The mechanism by which phagosomes associate with microtubules is not yet understood. One possibility is through members of the rab family of GTPases discussed above. Rab4a, rab5 and rab11 have been shown to associate with phagosomes and have demonstrated interactions with microtubules.[80-82]

Reorganization of the actin-based cytoskeleton is important during phagosomal maturation as well as phagosome formation. As in the case of microtubules described above, actin microfilaments appear to play a role in the centripetal movement of phagosome after their formation. While small phagosomes (approximately 1 μm in diabeter) can move along microtubules, a mechanism for phagosomal motility has been proposed that implicates myosin in the generation of force to move larger phagosomes along actin filaments.[74,83] It has also been proposed that assembly of actin filaments on the surface of the phagosome could generate motility.[84-86] In this model, members of the ezrin, radixin, moesin (ERM) family of proteins are described as regulators of actin nucleation that may initiate the formation of actin structures on the surface of phagosomes. This model is supported by the demonstration that ezrin and moesin can nucleate F-actin assembly on the surface of phagosomes in vitro.[85] F-actin assembly on the surface of endosomes, lysosomes and *Listeria* has been described as a force-generating mechanism in the movement of these intracellular compartments[87-89] and its role in phagosome motility in vivo needs to be investigated further.

In addition to supporting phagosomal motility, F-actin filaments may contribute to the maturation of the fully formed phagosome by facilitating the delivery of other endosomal compartments to the phagosome. Studies in an in vitro system suggest that an appropriate level of actin filament assembly is required to allow phagosome-endosome fusion. Enough actin filaments are needed, emanating from the phagosome, for the efficient transport of endosomes (presumably through the activity of myosin motors) to the phagosome.[86] An excess of F-actin around the phagosome, however, might impede the delivery of endosomes and retard biogenesis of the phagolysosome. One study suggests that stabilized F-actin filaments associated with *Leishmania*-containing phagosomes may contribute to the observed impairment in maturation in these phagosomes.[90] While much remains to be learned about this process in vivo, it is becoming clear that dynamic actin structures are prominent and functionally important features through the entire existence of a phagosome/phagolysosome.

Phosphoinositide Function in the Maturing Phagosome

Phosphoinositides play a critical role in phagosome formation, as discussed above, and in phagosomal maturation. Recent studies now indicate that, while class I PI3K functions to generate phosphatidylinositol 3,4,5-trisphosphate in a spatially restricted manner on the forming phagosome, it is class III PI3Ks (generating PI3P) that function during maturation of the phagosome.[91] Using GFP-tagged FYVE domains as a probe for PI3P production, Grinstein and coworkers have shown that soon after sealing of the nascent phagosome de novo synthesis of PI3P results in accumulation of PI3P on the phagosomal membrane.[91] This accumulation was wortmannin-sensitive and was inhibited by injection of antibodies to VPS34 into cells, but was not altered in cells deficient in the p85 subunit of class I PI3K. Importantly, in cells injected with VPS34 antibodies phagosomal maturation was prevented and revealed an impairment in the acquisition of LAMP-1 on the phagosome.[91] Thus, it is clear that generation of PI3P on phagosomes is required for maturation of the compartment and the mechanism by which this lipid facilitates this process is currently being studied.

Insight into the role of PI3P in phagosomal maturation may come from studies of the endocytic pathway. In this pathway, PI3P is important for the recruitment of EEA1 and other proteins containing FYVE domains or PX domains.[92,93] Microinjection of antibodies against EEA1 into macrophages impaired the delivery of late-endosomal markers to phagosomes, allowing speculation that EEA1 may function to physically link phagosomes to early endosomes.[94] The associations of other FYVE and PX domain-containing proteins, such as the p40[phox] and p47[phox] subunits of the NADPH oxidase, with maturing phagosomes remain to be characterized and a link to PI3P production has been proposed.[95] At the present time, our understanding of the function of PI3P and related phosphoinositides in phagocytosis is growing steadily. As our knowledge of the biochemical activities and biological functions of phosphoinositides, as well as other types of lipids increases, a fuller molecular understanding of phagolysosomal biogenesis comes within reach.

Intracellular Parasites and Inhibition of Maturation

Many intracellular pathogens have evolved strategies to overcome host defense mechanisms by interfering with phagosome maturation. Diverse strategies have been utilized, and space does not permit a complete survey, so below we outline how three organisms overcome the innate immune system to thrive within phagocytic cells.

Mycobacterium tuberculosis

Is the causative agent of tuberculosis and has been estimated to be carried by two billion people and cause over two million deaths per year worldwide. These Gram-positive rod-shaped bacteria of the actinomycete order survive within host macrophage cells by abrogating phagosome maturation. Interestingly, this inhibition of macrophage-mediated bacterial killing is dependent upon the viability of *M. tuberculosis*, emphasizing the dynamic nature of the host-pathogen relationship. Despite the importance of this pathogen and the large amount of research that has gone into its study, many controversies persist as to the cause and consequences of phagosomal maturation arrest. Phagosomes containing *M. tuberculosis* become arrested at the early phagosome stage, with accumulations of rab5 and a lack of rab7.[96] These phagosomes remain fusion competent, sustaining fusion events with early endosomal compartments.[97] Similar observations are seen with cells expressing constitutively active rab5, but whether the accumulated rab5 seen around *M. tuberculosis* phagosomes actively contributes to the arrest, or simply marks the last stage of development, remains to be determined. Rab5 recruitment and activation would be inconsistent with the recent observation of Deretic and colleagues who have found that latex beads coated with the bacterial lipid lipoarabinomannan (LAM), behave like *M. tuberculosis* and fail to mature.[98] LAM is structurally related to phosphatidylinositols, and LAM was found to inhibit the accumulation of EEA1 on phagosomes. While this could arrest maturation, it would be expected to inhibit the accumulation and activation of rab5, which is not observed for *M. tuberculosis* phagosomes. Also, the fact that dead bacteria cannot arrest maturation is inconsistent with maturation arrest seen with a lipid-coated latex bead. More recently, Deretic and coworkers have reported that a second cell wall lipid, phosphatidylinositol mannoside, stimulates early endosome fusion.[99]

Coronin, an actin-associated protein, was also found to concentrate on *M. tuberculosis* phagosomes[100] where it was proposed to act as a physical barrier to membrane fusion events. However, this explanation is somewhat inconsistent with observations pointed out above, that these phagosomes appear competent for fusion to early endosomes. Sequestration of cholesterol was also touted as a possible contributor to phagosome inhibition,[101] but since cholesterol depletion impairs complement receptor-mediated phagocytosis, the results of this study may have alternative explanations. Cytoplasmic Ca^{2+} elevations are normally seen following microbial phagocytosis and these have been shown to be impaired during *M. tuberculosis* phagocytosis.[102] Recently, Kusner's group have shown that *M. tuberculosis* inhibits sphingosine kinase, the enzyme responsible for phosphorylation of sphingosine to sphingosine-1-P, a lipid linked to elevation of intracellular Ca^{2+}.[103] Although it is not known how the kinase is inhibited, elevation of Ca^{2+} may be important for a number of signaling pathways leading to delivery of antimicrobial agents to the phagosome. Clearly, our understanding of maturation arrest mediated by *M. tuberculosis* remains limited and much additional work is needed.

Coxiella burnetii

Is a gram-negative coccobacillus that is prevalent around the world. It is an obligate intracellular pathogen that infects arthropods, birds and mammals. *C. burnetii* infection can cause coxiellosis in many domesticated animals such as cattle, sheep and goats and an atypical pneumonia, known as Q fever, in humans.[104-106] *C. burnetii* primarily targets monocytes and macrophages and is internalized by an incompletely described phagocytic process shown to involve $\alpha_V\beta_3$ integrin and $\alpha_M\beta_2$ integrin (CR3).[107] Once internalized, the *Coxiella* bacteria survive in unique phagolysosome-like compartments called large replicative vacuoles (LRVs), which can grow to a size that occupies much of the cytoplasm of the infected cell.[106,108] Formation of LRVs is the result of fusion of *Coxiella*-containing phagosomes with lysosomes and other endosomal compartments. The LRV compartment is acidic and maintenance of low pH is required for optimum *Coxiella* proliferation.[104,109] Consistent with their interaction with lysosomes, *Coxiella* LRVs contain lysosomal enzymes and membrane markers, such as LAMP-1.[109-111] Recent evidence indicates that Coxiella LRVs acquire monodansylcadaverine and the LC3 protein, two markers of autophagic vacuoles.[110] In this study it was also shown that pharmacological inhibition of autophagy impaired the formation of Coxiella LRVs. Interestingly, Beron et al report that *Coxiella* LRVs are rab7-positive and that expression of dominant-negative rab7 in *C. burnetii*-infected cells produces small *Coxiella*-containing LRVs.[110] These findings indicate that interaction with a rab7-containing endosomal compartment is required for maturation of the *C. burnetii* LRV, as has been described for the modified compartments generated by species of *Mycobacterium*, *Legionella* and *Salmonella*. It remains to be determined what components of the *Coxiella* LRV are dependent upon Rab7 function for their delivery to the LRV and how this might be linked the interaction with the autophagic pathway during biogenesis of the *C. burnetii* compartment.

Legionella pneumophila

Another important human pathogen that generates an intracellular niche in host cells is the gram-negative bacteria *Legionella pneumophila*. This pathogen can be inhaled in aerosol form and ingested by alveolar macrophages, in which it survives and replicates.[112,113] Within macrophages, phagosomes containing *L. pneumophila* escape normal maturation; they do not acquire markers of late endosomes (save for rab7, see below) or lysosomes and they do not acidify.[114-116] Rather, within 4 hours of ingestion, *Legionella*-containing phagosomes connect with the macrophage endoplasmic reticulum.[117,118] This results in a unique, ribosome-bearing compartment that supports *Legionella* replication. A 4-stage model has been proposed for this organelle biogenesis that is induced by *L. pneumophila*.[118] This model involves (1) inhibition of phagosomal fusion with early endosomes, (2) attachment of ER vesicles and mitochondria to the phagosome, (3) alteration of the phagosomal membrane thickness to resemble that of ER vesicles and (4) attachment of ribosomes directly to the *Legionella*-containing compartment. By the 4th stage the bacteria are in a compartment that is biochemically very similar, if not identical, to the host cell ER. How the pathogen-bearing phagosome establishes interactions with host organelles such as the ER is currently being studied and it is at least speculated that attachment of ribosomes may be effected by fusion with rough ER-derived vesicles.[118]

The molecular activities that control the intracellular survival of *L. pneumophila* have not been completely characterized, but stages 1 and 2 of the compartmental biogenesis appear to be dependent on the Dot/Icm (defective organelle trafficking/intracellular multiplication) type IV secretion system.[118-120] This secretion system is thought to deliver into the host cell cytosol factors that permit association of specific vesicles with the phagosomes and a separate set of factors that inhibit interaction between the *Legionella* compartment and late endosomes or lysosomes.[112,118] Recently, one of these factors, RalF, was identified as a guanine nucleotide exchange factor (GEF) for ADP ribosylation factor (ARF).[120] RalF was shown to be required for recruitment of ARF to the *Legionella* phagosome, but it was not required for the inhibition of fusion with lysosomes, fusion with the ER or intracellular replication of *Legionella*.[120] There is currently much speculation as to the nature of additional virulence factors that may be injected by *Legionella* into the host cell cytosol. It has been suggested that injected *Legionella* proteins might function as regulators of membrane traffic, such as SNAREs or tethering proteins, to induce interactions with host cell ER vesicles.[112] *Legionella* phagosomes transiently bind rab7,[121] which is known to regulate fusion events involving late endosomes, but the physiological significance of this interaction is not clear. It has been demonstrated, using a HeLa cell model system, that recruitment of rab7 to the surface of *Legionella*-containing phagosomes does not promote maturation of the phagosome, at least within 8 hours of internalization.[114] It is possible that the effects of rab7 activity may only be apparent at a later stage in *Legionella* infection and one study does indicate that *Legionella* compartments can acquire lysosomal characteristics, including acidification, 18 hours after ingestion of the bacteria.[122] Together, these findings suggest that biogenesis of the *Legionella* compartment may yet be more complicated than first conceived. The identification of additional factors injected by *Legionella* will be critical to our understanding of this pathogen.

Summary

The progressive maturation of the phagosome is an essential process in the killing of microbial pathogens, yet many of the steps involved remain obscure. Importantly, many pathogenic microbes have developed means to perturb cellular processes in order to avoid destruction. Through the further study of the basic mechanisms controlling phagosome formation and maturation, and through the identification of the steps inhibited by microbes, the development of novel therapeutic strategies will be possible to treat some of mankind's most deadly illnesses.

References

1. Bonifacino J, Glick B. The mechanism of vesicle budding and fusion. Cell 2004; 116:153-166.
2. Rothman JE, Warren G. Implications of the SNARE hypothesis for intracellular membrane topology and dynamics. Curr Biol 1994; 4(3):220-233.
3. Trimble WS, Cowan DM, Scheller RH. VAMP-1: A synaptic vesicle-associated integral membrane protein. Proc Natl Acad Sci USA 1988; 85(12):4538-4542.
4. Bennett MK, Calakos N, Scheller RH. Syntaxin: A synaptic protein implicated in docking of synaptic vesicles at presynaptic active zones. Science. 1992; 257(5067):255-259.
5. Oyler GA, Higgins GA, Hart RA et al. The identification of a novel synaptosomal-associated protein, SNAP-25, differentially expressed by neuronal subpopulations. J Cell Biol 1989; 109(6 Pt 1):3039-3052.
6. Sollner T, Bennett M, Whiteheart S et al. A protein assembly-disassembly pathway in vitro that may correspond to sequential steps of synaptic vesicle docking, activation, and fusion. Cell 1993; 75(3):409-418.
7. Block M, Glick B, Wilcox C et al. Purification of an N-ethylmaleimide-sensitive protein catalyzing vesicular transport. Proc Natl Acad Sci USA 1988; 85(21):7852-7856.
8. Clary DO, Griff IC, Rothman JE. SNAPs, a family of NSF attachment proteins involved in intracellular membrane fusion in animals and yeast. Cell 1990; 61(4):709-721.
9. Weber T, Zemelman B, McNew J et al. SNAREpins: Minimal machinery for membrane fusion. Cell 1998; 92:759-772.
10. Bock JB, Matern HT, Peden AA et al. A genomic perspective on membrane compartment organization. Nature 2001; 409(6822):839-841.
11. Fasshauer D, Antonin W, Margittai M et al. Mixed and noncognate SNARE complexes. Characterization of assembly and biophysical properties. J Biol Chem 1999; 274(22):15440-15446.
12. McNew J, Parlati F, Fukuda R et al. Compartmental specificity of cellular membrane fusion encoded in SNARE proteins. Nature 2000; 407:153-159.
13. Chen YA, Scheller RH. SNARE-mediated membrane fusion. Nat Rev Mol Cell Biol 2001; 2(2):98-106.
14. Galli T, Chilcote T, Mundigl O et al. Tetanus toxin-mediated cleavage of cellubrevin impairs exocytosis of transferrin receptor-containing vesicles in CHO cells. J Cell Biol 1994; 125(5):1015-1024.
15. Foran PG, Fletcher LM, Oatey PB et al. Protein kinase B stimulates the translocation of GLUT4 but not GLUT1 or transferrin receptors in 3T3-L1 adipocytes by a pathway involving SNAP-23, synaptobrevin-2, and/or cellubrevin. J Biol Chem 1999; 274(40):28087-28095.
16. Mullock BM, Smith CW, Ihrke G et al. Syntaxin 7 is localized to late endosome compartments, associates with Vamp 8, and Is required for late endosome-lysosome fusion. Mol Biol Cell 2000; 11(9):3137-3153.
17. Deneka M, Neeft M, van der Sluijs P. Regulation of membrane transport by rab GTPases. Crit Rev Biochem Mol Biol 2003; 38(2):121-142.
18. Birkeland HC, Stenmark H. Protein targeting to endosomes and phagosomes via FYVE and PX domains. Curr Top Microbiol Immunol 2004; 282:89-115.
19. Greenberg S, Chang P, Wang DC et al. Clustered syk tyrosine kinase domains trigger phagocytosis. Proc Natl Acad Sci USA 1996; 93(3):1103-1107.
20. Crowley MT, Costello PS, Fitzer-Attas CJ et al. A critical role for Syk in signal transduction and phagocytosis mediated by Fcgamma receptors on macrophages. J Exp Med 1997; 186(7):1027-1039.

21. Matsuda M, Park JG, Wang DC et al. Abrogation of the Fc gamma receptor IIA-mediated phago-cytic signal by stem-loop Syk antisense oligonucleotides. Mol Biol Cell 1996; 7(7):1095-1106.
22. Kiefer F, Brumell J, Al-Alawi N et al. The Syk protein tyrosine kinase is essential for Fcgamma receptor signaling in macrophages and neutrophils. Mol Cell Biol 1998; 18(7):4209-4220.
23. Oliver JM, Burg DL, Wilson BS et al. Inhibition of mast cell Fc epsilon R1-mediated signaling and effector function by the Syk-selective inhibitor, piceatannol. J Biol Chem 1994; 269(47):29697-29703.
24. Coppolino MG, Krause M, Hagendorff P et al. Evidence for a molecular complex consisting of Fyb/SLAP, SLP-76, Nck, VASP and WASP that links the actin cytoskeleton to Fcgamma receptor signalling during phagocytosis. J Cell Sci 2001; 114(Pt 23):4307-4318.
25. Machesky LM, Gould KL. The Arp2/3 complex: A multifunctional actin organizer. Curr Opin Cell Biol 1999; 11(1):117-121.
26. Machesky LM, Insall RH. Signaling to actin dynamics. J Cell Biol 1999; 146(2):267-272.
27. May RC, Caron E, Hall A et al. Involvement of the Arp2/3 complex in phagocytosis mediated by FcgammaR or CR3. Nat Cell Biol 2000; 2(4):246-248.
28. Machesky LM, Insall RH. Scar1 and the related Wiskott-Aldrich syndrome protein, WASP, regu-late the actin cytoskeleton through the Arp2/3 complex. Curr Biol 1998; 8(25):1347-1356.
29. Lorenzi R, Brickell PM, Katz DR et al. Wiskott-Aldrich syndrome protein is necessary for efficient IgG-mediated phagocytosis. Blood 2000; 95(9):2943-2946.
30. Rohatgi R, Ma L, Miki H et al. The interaction between N-WASP and the Arp2/3 complex links Cdc42-dependent signals to actin assembly. Cell 1999; 97(2):221-231.
31. Serrander L, Skarman P, Rasmussen B et al. Selective inhibition of IgG-mediated phagocytosis in gelsolin-deficient murine neutrophils. J Immunol 2000; 165(5):2451-2457.
32. Aizawa H, Fukui Y, Yahara I. Live dynamics of Dictyostelium cofilin suggests a role in remodeling actin latticework into bundles. J Cell Sci 1997; 110(Pt 19):2333-2344.
33. Nagaishi K, Adachi R, Matsui S et al. Herbimycin a inhibits both dephosphorylation and translo-cation of cofilin induced by opsonized zymosan in macrophagelike U937 cells. J Cell Physiol 1999; 180(3):345-354.
34. Allen LA, Aderem A. Molecular definition of distinct cytoskeletal structures involved in comple-ment- and Fc receptor-mediated phagocytosis in macrophages. J Exp Med 1996; 184(2):627-637.
35. Rivero F, Furukawa R, Fechheimer M et al. Three actin cross-linking proteins, the 34 kDa actin-bundling protein, alpha-actinin and gelation factor (ABP-120), have both unique and redun-dant roles in the growth and development of Dictyostelium. J Cell Sci 1999; 112(Pt 16):2737-2751.
36. Cox D, Wessels D, Soll DR et al. Reexpression of ABP-120 rescues cytoskeletal, motility, and phagocytosis defects of ABP-120- Dictyostelium mutants. Mol Biol Cell 1996; 7(5):803-823.
37. Rivero F, Koppel B, Peracino B et al. The role of the cortical cytoskeleton: F-actin crosslinking proteins protect against osmotic stress, ensure cell size, cell shape and motility, and contribute to phagocytosis and development. J Cell Sci 1996; 109(Pt 11):2679-2691.
38. Allen LA, Aderem A. Protein kinase C regulates MARCKS cycling between the plasma membrane and lysosomes in fibroblasts. EMBO J 1995; 14(6):1109-1120.
39. Carballo E, Pitterle DM, Stumpo DJ et al. Phagocytic and macropinocytic activity in MARCKS-deficient macrophages and fibroblasts. Am J Physiol 1999; 277(1 Pt 1):C163-173.
40. Greenberg S, Burridge K, Silverstein SC. Colocalization of F-actin and talin during Fc receptor-mediated phagocytosis in mouse macrophages. J Exp Med 1990; 172(6):1853-1856.
41. Greenberg S, Chang P, Silverstein SC. Tyrosine phosphorylation of the gamma subunit of Fc gamma receptors, p72syk, and paxillin during Fc receptor-mediated phagocytosis in macrophages. J Biol Chem 1994; 269(5):3897-3902.
42. Pan XQ, Darby C, Indik ZK et al. Activation of three classes of nonreceptor tyrosine kinases following Fc gamma receptor crosslinking in human monocytes. Clin Immunol 1999; 90(1):55-64.
43. Botelho RJ, Teruel M, Dierckman R et al. Localized biphasic changes in phosphatidylinositol-4,5-bisphosphate at sites of phagocytosis. J Cell Biol 2000; 151(7):1353-1368.
44. Coppolino MG, Dierckman R, Loijens J et al. Inhibition of phosphatidylinositol-4-phosphate 5-ki-nase Ialpha impairs localized actin remodeling and suppresses phagocytosis. J Biol Chem 2002; 277(46):43849-43857.

45. Lowry MB, Duchemin AM, Robinson JM et al. Functional separation of pseudopod extension and particle internalization during Fc gamma receptor-mediated phagocytosis. J Exp Med 1998; 187(2):161-176.

46. Cox D, Tseng CC, Bjekic G et al. A requirement for phosphatidylinositol 3-kinase in pseudopod extension. J Biol Chem 1999; 274(3):1240-1247.

47. Araki N, Johnson MT, Swanson JA. A role for phosphoinositide 3-kinase in the completion of macropinocytosis and phagocytosis by macrophages. J Cell Biol 1996; 135(5):1249-1260.

48. Werb Z, Cohn ZA. Plasma membrane synthesis in the macrophage following phagocytosis of polystyrene latex particles. J Biol Chem 1972; 247(8):2439-2446.

49. Holevinsky KO, Nelson DJ. Membrane capacitance changes associated with particle uptake during phagocytosis in macrophages. Biophys J 1998; 75(5):2577-2586.

50. Hackam DJ, Rotstein OD, Sjolin C et al. v-SNARE-dependent secretion is required for phagocytosis. Proc Natl Acad Sci USA 1998; 95(20):11691-11696.

51. Coppolino MG, Kong C, Mohtashami M et al. Requirement for N-ethylmaleimide-sensitive factor activity at different stages of bacterial invasion and phagocytosis. J Biol Chem 2001; 276(7):4772-4780.

52. Bajno L, Peng XR, Schreiber AD et al. Focal exocytosis of VAMP3-containing vesicles at sites of phagosome formation [see comments]. J Cell Biol 2000; 149(3):697-706.

53. Niedergang F, Colucci-Guyon E, Dubois T et al. ADP ribosylation factor 6 is activated and controls membrane delivery during phagocytosis in macrophages. J Cell Biol 2003; 161(6):1143-1150.

54. Allen LA, Yang C, Pessin JE. Rate and extent of phagocytosis in macrophages lacking vamp3. J Leukoc Biol 2002; 72(1):217-221.

55. Gagnon E, Duclos S, Rondeau C et al. Endoplasmic reticulum-mediated phagocytosis is a mechanism of entry into macrophages. Cell 2002; 110(1):119-131.

56. Berg JS, Cheney RE. Myosin-X is an unconventional myosin that undergoes intrafilopodial motility. Nat Cell Biol 2002; 4(3):246-250.

57. Cox D, Berg JS, Cammer M et al. Myosin X is a downstream effector of PI(3)K during phagocytosis. Nat Cell Biol 2002; 4(7):469-477.

58. Araki N, Hatae T, Furukawa A et al. Phosphoinositide-3-kinase-independent contractile activities associated with Fcgamma-receptor-mediated phagocytosis and macropinocytosis in macrophages. J Cell Sci 2003; 116(Pt 2):247-257.

59. Tjelle TE, Lovdal T, Berg T. Phagosome dynamics and function. Bioessays 2000; 22(3):255-263.

60. Mayorga LS, Bertini F, Stahl PD. Fusion of newly formed phagosomes with endosomes in intact cells and in a cell-free system. J Biol Chem 1991; 266(10):6511-6517.

61. Funato K, Beron W, Yang CZ et al. Reconstitution of phagosome-lysosome fusion in streptolysin O- permeabilized cells. J Biol Chem 1997; 272(26):16147-16151.

62. Claus V, Jahraus A, Tjelle T et al. Lysosomal enzyme trafficking between phagosomes, endosomes, and lysosomes in J774 macrophages. Enrichment of cathepsin H in early endosomes. J Biol Chem 1998; 273(16):9842-9851.

63. Jahraus A, Tjelle TE, Berg T et al. In vitro fusion of phagosomes with different endocytic organelles from J774 macrophages. J Biol Chem 1998; 273(46):30379-30390.

64. Alvarez-Dominguez C, Barbieri AM, Beron W et al. Phagocytosed live listeria monocytogenes influences Rab5-regulated in vitro phagosome-endosome fusion. J Biol Chem 1996; 271(23):13834-13843.

65. Vieira OV, Bucci C, Harrison RE et al. Modulation of Rab5 and Rab7 recruitment to phagosomes by phosphatidylinositol 3-kinase. Mol Cell Biol 2003; 23(7):2501-2514.

66. Duclos S, Diez R, Garin J et al. Rab5 regulates the kiss and run fusion between phagosomes and endosomes and the acquisition of phagosome leishmanicidal properties in RAW 264.7 macrophages. J Cell Sci 2000; 113(Pt 19):3531-3541.

67. Feng Y, Press B, Wandinger-Ness A. Rab 7: An important regulator of late endocytic membrane traffic. J Cell Biol 1995; 131(6 Pt 1):1435-1452.

68. Vitelli R, Santillo M, Lattero D et al. Role of the small GTPase Rab7 in the late endocytic pathway. J Biol Chem 1997; 272(7):4391-4397.

69. Rupper A, Grove B, Cardelli J. Rab7 regulates phagosome maturation in Dictyostelium. J Cell Sci 2001; 114(Pt 13):2449-2460.

70. Desjardins M. Biogenesis of phagolysosomes: The 'kiss and run' hypothesis. Trends Cell Biol 1995; 5(5):183-186.

71. Tse SM, Furuya W, Gold E et al. Differential role of actin, clathrin, and dynamin in Fc gamma receptor-mediated endocytosis and phagocytosis. J Biol Chem 2003; 278(5):3331-3338.

72. Booth JW, Kim MK, Jankowski A et al. Contrasting requirements for ubiquitylation during Fc receptor-mediated endocytosis and phagocytosis. EMBO J 2002; 21(3):251-258.

73. Botelho RJ, Hackam DJ, Schreiber AD et al. Role of COPI in phagosome maturation. J Biol Chem 2000; 275(21):15717-15727.

74. Toyohara A, Inaba K. Transport of phagosomes in mouse peritoneal macrophages. J Cell Sci 1989; 94(Pt 1):143-153.

75. Blocker A, Severin FF, Burkhardt JK et al. Molecular requirements for bi-directional movement of phagosomes along microtubules. J Cell Biol 1997; 137(1):113-129.

76. Blocker A, Griffiths G, Olivo JC et al. A role for microtubule dynamics in phagosome movement. J Cell Sci 1998; 111(Pt 3):303-312.

77. Desjardins M, Huber LA, Parton RG et al. Biogenesis of phagolysosomes proceeds through a sequential series of interactions with the endocytic apparatus. J Cell Biol 1994; 124(5):677-688.

78. Blocker A, Severin FF, Habermann A et al. Microtubule-associated protein-dependent binding of phagosomes to microtubules. J Biol Chem 1996; 271(7):3803-3811.

79. Funato K, Beron W, Yang CZ et al. Reconstitution of phagosome-lysosome fusion in streptolysin O-permeabilized cells. J Biol Chem 1997; 272(26):16147-16151.

80. Nielsen E, Severin F, Backer JM et al. Rab5 regulates motility of early endosomes on microtubules. Nat Cell Biol 1999; 1(6):376-382.

81. Mammoto A, Ohtsuka T, Hotta I et al. Rab11BP/Rabphilin-11, a downstream target of rab11 small G protein implicated in vesicle recycling. J Biol Chem 1999; 274(36):25517-25524.

82. Bielli A, Thornqvist PO, Hendrick AG et al. The small GTPase Rab4A interacts with the central region of cytoplasmic dynein light intermediate chain-1. Biochem Biophys Res Commun 2001; 281(5):1141-1153.

83. Moller W, Nemoto I, Matsuzaki T et al. Magnetic phagosome motion in J774A.1 macrophages: Influence of cytoskeletal drugs. Biophys J 2000; 79(2):720-730.

84. Defacque H, Egeberg M, Antzberger A et al. Actin assembly induced by polylysine beads or purified phagosomes: Quantitation by a new flow cytometry assay. Cytometry 2000; 41(1):46-54.

85. Defacque H, Egeberg M, Habermann A et al. Involvement of ezrin/moesin in de novo actin assembly on phagosomal membranes. EMBO J 2000; 19(2):199-212.

86. Jahraus A, Egeberg M, Hinner B et al. ATP-dependent membrane assembly of F-actin facilitates membrane fusion. Mol Biol Cell 2001; 12(1):155-170.

87. Taunton J, Rowning BA, Coughlin ML et al. Actin-dependent propulsion of endosomes and lysosomes by recruitment of N-WASP. J Cell Biol 2000; 148(3):519-530.

88. Taunton J. Actin filament nucleation by endosomes, lysosomes and secretory vesicles. Curr Opin Cell Biol 2001; 13(1):85-91.

89. Tilney LG, Connelly PS, Portnoy DA. Actin filament nucleation by the bacterial pathogen, Listeria monocytogenes. J Cell Biol 1990; 111(6 Pt 2):2979-2988.

90. Holm A, Tejle K, Magnusson KE et al. Leishmania donovani lipophosphoglycan causes periphagosomal actin accumulation: Correlation with impaired translocation of PKCalpha and defective phagosome maturation. Cell Microbiol 2001; 3(7):439-447.

91. Vieira OV, Botelho RJ, Rameh L et al. Distinct roles of class I and class III phosphatidylinositol 3-kinases in phagosome formation and maturation. J Cell Biol 2001; 155(1):19-25.

92. Gillooly DJ, Raiborg C, Stenmark H. Phosphatidylinositol 3-phosphate is found in microdomains of early endosomes. Histochem Cell Biol 2003; 120(6):445-453.

93. Tuma PL, Nyasae LK, Backer JM et al. Vps34p differentially regulates endocytosis from the apical and basolateral domains in polarized hepatic cells. J Cell Biol 2001; 154(6):1197-1208.

94. Fratti RA, Backer JM, Gruenberg J et al. Role of phosphatidylinositol 3-kinase and Rab5 effectors in phagosomal biogenesis and mycobacterial phagosome maturation arrest. J Cell Biol 2001; 154(3):631-644.

95. Kanai F, Liu H, Field SJ et al. The PX domains of p47phox and p40phox bind to lipid products of PI(3)K. Nat Cell Biol 2001; 3(7):675-678.

96. Via LE, Deretic D, Ulmer RJ et al. Arrest of mycobacterial phagosome maturation is caused by a block in vesicle fusion between stages controlled by rab5 and rab7. J Biol Chem 1997; 272(20):13326-13331.

97. Sturgill-Koszycki S, Schaible UE, Russell DG. Mycobacterium-containing phagosomes are accessible to early endosomes and reflect a transitional state in normal phagosome biogenesis. EMBO J 1996; 15(24):6960-6968.

98. Fratti RA, Chua J, Vergne I et al. Mycobacterium tuberculosis glycosylated phosphatidylinositol causes phagosome maturation arrest. Proc Natl Acad Sci USA 2003; 100(9):5437-5442.

99. Vergne I, Fratti RA, Hill PJ et al. Mycobacterium tuberculosis phagosome maturation arrest: Mycobacterial phosphatidylinositol analog PIM stimulates early endosomal fusion. Mol Biol Cell 2004; 12(2):751-760.

100. Ferrari G, Langen H, Naito M et al. A coat protein on phagosomes involved in the intracellular survival of mycobacteria. Cell 1999; 97(4):435-447.

101. Gatfield J, Pieters J. Essential role for cholesterol in entry of mycobacteria into macrophages. Science 2000; 288(5471):1647-1650.

102. Malik ZA, Iyer SS, Kusner DJ. Mycobacterium tuberculosis phagosomes exhibit altered calmodulin-dependent signal transduction: Contribution to inhibition of phagosome-lysosome fusion and intracellular survival in human macrophages. J Immunol 2001; 166(5):3392-3401.

103. Malik ZA, Thompson CR, Hashimi S et al. Cutting edge: Mycobacterium tuberculosis blocks Ca2+ signaling and phagosome maturation in human macrophages via specific inhibition of sphingosine kinase. J Immunol 2003; 170(6):2811-2815.

104. Hackstadt T, Williams JC. Biochemical stratagem for obligate parasitism of eukaryotic cells by Coxiella burnetii. Proc Natl Acad Sci USA 1981; 78(5):3240-3244.

105. Marrie TJ. Coxiella burnetii pneumonia. Eur Respir J 2003; 21(4):713-719.

106. Norlander L. Q fever epidemiology and pathogenesis. Microbes Infect 2000; 2(4):417-424.

107. Capo C, Lindberg FP, Meconi S et al. Subversion of monocyte functions by coxiella burnetii: Impairment of the cross-talk between alphavbeta3 integrin and CR3. J Immunol 1999; 163(11):6078-6085.

108. Baca OG, Paretsky D. Q fever and Coxiella burnetii: A model for host-parasite interactions. Microbiol Rev 1983; 47(2):127-149.

109. Heinzen RA, Scidmore MA, Rockey DD et al. Differential interaction with endocytic and exocytic pathways distinguish parasitophorous vacuoles of Coxiella burnetii and Chlamydia trachomatis. Infect Immun 1996; 64(3):796-809.

110. Beron W, Gutierrez MG, Rabinovitch M et al. Coxiella burnetii localizes in a Rab7-labeled compartment with autophagic characteristics. Infect Immun 2002; 70(10):5816-5821.

111. Lem L, Riethof DA, ScidmoreCarlson M et al. Enhanced interaction of HLA-DM with HLA-DR in enlarged vacuoles of hereditary and infectious lysosomal diseases. J Immunol 1999; 162(1):523-532.

112. Roy CR, Tilney LG. The road less traveled: Transport of Legionella to the endoplasmic reticulum. J Cell Biol 2002; 158(3):415-419.

113. Horwitz MA, Silverstein SC. Legionnaires' disease bacterium (Legionella pneumophila) multiples intracellularly in human monocytes. J Clin Invest 1980; 66(3):441-450.

114. Clemens DL, Lee BY, Horwitz MA. Mycobacterium tuberculosis and Legionella pneumophila phagosomes exhibit arrested maturation despite acquisition of Rab7. Infect Immun 2000; 68(9):5154-5166.

115. Horwitz MA. The Legionnaires' disease bacterium (Legionella pneumophila) inhibits phagosome-lysosome fusion in human monocytes. J Exp Med 1983; 158(6):2108-2126.

116. Horwitz MA, Maxfield FR. Legionella pneumophila inhibits acidification of its phagosome in human monocytes. J Cell Biol 1984; 99(6):1936-1943.

117. Swanson MS, Isberg RR. Association of Legionella pneumophila with the macrophage endoplasmic reticulum. Infect Immun 1995; 63(9):3609-3620.

118. Tilney LG, Harb OS, Connelly PS et al. How the parasitic bacterium Legionella pneumophila modifies its phagosome and transforms it into rough ER: Implications for conversion of plasma membrane to the ER membrane. J Cell Sci 2001; 114(Pt 24):4637-4650.

119. Nagai H, Roy CR. The DotA protein from Legionella pneumophila is secreted by a novel process that requires the Dot/Icm transporter. EMBO J 2001; 20(21):5962-5970.

120. Nagai H, Kagan JC, Zhu X et al. A bacterial guanine nucleotide exchange factor activates ARF on Legionella phagosomes. Science 2002; 295(5555):679-682.

121. Roy CR, Berger KH, Isberg RR. Legionella pneumophila DotA protein is required for early phagosome trafficking decisions that occur within minutes of bacterial uptake. Mol Microbiol 1998; 28(3):663-674.

122. Sturgill-Koszycki S, Swanson MS. Legionella pneumophila replication vacuoles mature into acidic, endocytic organelles J Exp Med 2000; 192(9):1261-1272.

Index

A

Actin 2, 7, 8, 12-17, 25, 35-39, 54, 72-76, 78, 79, 87, 97, 98, 103, 105, 110, 117-120, 123-127, 136-139, 141-143
Actin-binding protein 17, 123, 126
Actin cytoskeleton 12, 37-39, 54, 74, 76, 79, 117, 118, 120
Alternative pathway 50, 53
Apoptosis 3, 12, 66, 89, 120
Apoptotic cell (AC) 1-12, 14-17, 24, 26-28, 59, 60, 62-64, 66, 73, 74, 77, 118, 120
Arachidonic acid 7, 10-12, 37, 64, 65, 67, 99, 106, 111, 121
Arf 72, 78, 79
Arf6 72, 78, 79
Arp2/3 2, 13-16, 36, 54, 74, 76, 79, 136
Autoimmunity 23, 27, 28

B

B cell 5, 27, 28, 86-91
Bordetella 75
Brucella abortus 75

C

C2 domain 91, 100, 101, 107, 111
C3 49-53, 55
Calcium signaling 2, 25, 37, 40, 117, 119, 121, 127
CD11b 24, 51, 74, 119
CD11c 24, 51, 119
CD18 24, 51, 52, 74, 119, 141
CD35 24, 51, 119
CD36 4, 5, 7, 9, 24
CD59 5
CD91 4, 6-8, 27
Cell activation 3, 33, 53, 54, 62
Cell differentiation 62
Ceramide 11, 99, 102, 104, 105, 108-111
Classical pathway 49, 50
Clostridium 73, 75

Complement

Complement 1-4, 6, 8, 11, 12, 14, 23, 24, 28, 39, 40, 49-51, 53-55, 59, 60, 62-64, 66, 74, 76, 85, 97, 98, 103, 108, 118-120, 125, 136, 143
Complement receptor (CR) 2-4, 6-8, 10-14, 16, 23, 28, 40, 49-51, 53-55, 59, 62, 74, 85, 98, 103, 118, 119, 136, 143
  CR1 3, 24, 51, 97, 119
  CR3 3, 4, 10, 14, 24, 39, 40, 51, 54, 55, 59, 60, 74, 76, 77, 79, 97, 119, 120, 123, 125, 136, 143
  CR4 3, 4, 6, 7, 24, 51, 59, 60, 119
Coxiella burnetii 143
C-reactive protein 28, 34
Cross-priming 26-28
Cytokine 2, 12, 26, 54, 59-62, 66, 85, 87-90, 119
Cytoskeleton 1, 2, 8, 12-15, 17, 37-39, 54, 55, 74-76, 79, 117, 118, 120, 127, 141

D

Degranulation 58, 59, 66, 88, 89, 106
Dendritic cell 1, 4, 7, 23, 24, 26, 58, 97
Diacylglycerol (DAG) 7-9, 11, 37, 61, 99-103, 105, 111, 112, 121, 125, 137
Dictyostelium 73, 74, 76-78, 117, 137, 140
Dynamin 39, 141

E

Early endosome antigen 1 (EEA1) 39, 98, 110, 126, 135, 139, 140, 142, 143
Endocytic pathway 125, 127, 133, 135, 140, 141, 142
Endoplasmic reticulum (ER) 6, 8-10, 25, 26, 37, 51, 78, 120-123, 126, 138, 144
Endosomes 26, 39, 78, 79, 98, 105, 106, 110, 120, 123, 125-127, 133-135, 138-144
Escherichia coli 24, 75-77
Extracellular signal regulated kinase (ERK) 7-9, 11, 12, 25, 36-38, 61-65, 98, 101, 102, 106, 108-110

F

Fc receptor (FcR) 2, 3, 6, 7, 11, 13, 33, 34, 35, 36, 38, 39, 40, 53, 55, 59, 73, 85, 119, 120, 133, 136, 138, 141
Fcγ receptor (FcγR) 2, 6, 8-16, 23-25, 27, 33-40, 55, 59, 60, 62, 64, 73, 74, 77, 78, 85-92, 97, 98, 100-103, 105, 106, 109, 110, 118-120, 123-125, 127, 136, 137, 141
  FcγRI 2, 24, 33, 34, 40, 64, 85, 86, 89, 100, 119, 123, 124, 127
  FcγRII 2, 33, 85, 88, 127
  FcγRIII 2, 24, 33, 34, 85
Fibroblast 1, 58, 77, 88, 91, 105
Focal adhesion kinase (FAK) 25, 91, 137

G

Gelsolin 37, 38, 110, 124, 125, 136, 137
GM-CSF 54
GTP-binding protein 72-74, 76, 77-80
GTPase 2, 10, 12-15, 17, 25, 36, 54, 72, 73, 75-80, 87, 125, 135, 136, 141
Guanine nucleotide-exchange factor (GEF) 9, 13, 14, 36, 72-74, 77, 87, 144
GULP 7

H

Haemophilus 25
Host defense 1, 16, 49, 58, 67, 142

I

IgG 2, 3, 23, 24, 27, 33, 34, 39, 40, 53-55, 59, 61, 62, 64, 65, 77, 78, 87, 88, 92, 97, 98, 100-103, 105-110, 118-120, 123, 124, 139
IL-4 54, 88
Immunoglobulin (Ig) 1-3, 33, 59-61, 63-66, 73, 97, 100, 119, 136
Immunoreceptor tyrosine based activation motif (ITAM) 6, 7, 24, 25, 34, 35, 73, 85-88, 90, 92, 100-102, 111, 120, 124
Immunoreceptor tyrosine-based inhibitory motif (ITIM) 27, 34, 40, 87-90
Inside-out signaling 52, 53

Integrin 3-6, 10-12, 14, 24, 25, 39, 40, 51-56, 75-77, 79, 118, 119, 122, 143
Intracellular parasite 133, 142

K

Kupffer cell 97

L

Langerhans cell 97
Legionella pneumophila 25, 138, 144, 149, 150
Leishmania amazonensis 75
Lens epithelial cell 1
Leukotrienes 25, 59, 62, 67, 108, 121
Lipid kinase 24, 110, 136
Listeria monocytogenes 75
Lysophospholipid 106, 108
Lysosomes 23, 25, 26, 37-39, 76, 98, 123, 125, 126, 133, 135, 139-141, 143, 144

M

M-CSF 54
MacMARCKS 125
Macrophage 1-7, 10-12, 16, 24, 27, 33, 34, 37, 54, 55, 58, 59, 61-66, 72, 73, 76-78, 85-92, 97, 98, 100-103, 105-109, 111, 118-120, 123, 125, 126, 133, 136, 138, 142-144
Mannose binding protein (MBP) 50
MARCKS 125, 137
MASP-1 50
MBP pathway 50
MER 4, 6, 7, 9, 59
MHC Class I 5, 26
MHC Class II 25, 26, 28
Microglia 1, 4, 58, 97
Microtubule 12, 14, 75, 76, 126, 141
Monocyte 1-3, 24, 33, 37, 58, 60-66, 87, 88, 90, 97, 100, 102, 103, 106, 111, 123, 143
Moraxella 25
Mycobacterium tuberculosis 7, 24, 25, 98, 103, 118, 142, 143
Myosin 2, 7, 8, 12-17, 25, 38, 39, 54, 75, 118, 138, 139, 141, 142

N

NADPH oxidase 79, 101, 102, 105, 107-111, 142
Natural killer (NK) cell 2, 88
Neisseria 25, 73
Neutrophil 1-3, 10-12, 25, 33, 34, 37, 40, 55, 58, 61, 62, 66, 97, 100-103, 108, 110, 111, 118, 119, 123, 125, 137, 138
N. meningitidis 75
Nonprofessional phagocyte 58, 59, 62, 66, 118

O

Opsonin 1-3, 17, 27, 28, 34, 53, 55, 66, 67, 75, 97, 118, 119
Opsonization 49-51, 53, 55, 66, 119

P

PAMP receptor 50
Paxillin 137
Phagocytic cup 35-37, 39, 77, 123, 126, 139, 140
Phagocytosis 1-17, 23-28, 33-40, 49, 51, 53-55, 58-66, 72-80, 85-92, 97-111, 117-120, 123-127, 133, 136-143
Phagolysosome 37, 38, 72, 125, 126, 133, 141-143
Phagolysosome fusion 37, 38
Phagosome 2, 7-17, 25, 26, 35, 37-39, 66, 74, 75, 77-79, 97, 98, 100-106, 109, 110, 117, 118, 120, 123-127, 133, 134, 136-145
Phosphatidylinositol 2, 6-9, 11, 24, 27, 33, 40, 61-63, 65, 76, 99, 111, 122, 124, 135-137, 142, 143
Phosphatidylinositol 3-kinase (PI 3-K) 6-10, 24, 25, 27, 36-39, 60-65, 100, 101, 103, 104, 110, 135, 137, 138, 140, 142
Phosphatidylinositol phosphate (PIP) 137
Phosphatidylserine receptor (PSR) 4-6, 9, 10, 24, 27, 28, 59, 60
Phospholipase 1, 2, 6-11, 17, 27, 35-37, 61, 62, 64, 65, 97-99, 102, 105, 106, 108, 111, 112, 122-125, 137
Phospholipase A2 (PLA2) 7-12, 25, 61, 64, 65, 99, 103, 105-108, 111, 112
Phospholipase Cγ (PLCγ) 7, 9, 11, 37, 61, 86, 100-102, 110, 122-124, 137

Phospholipase D (PLD) 7, 8, 10, 11, 25, 37, 61, 99, 101-105, 108-112, 123, 124
Phospholipid 4, 24, 86, 90, 97-99, 106, 107, 111, 112, 118, 120, 123
Platelets 66, 125
Plekstrin 125
Professional phagocyte 1, 2, 5, 16, 49, 58, 59, 61, 62, 64-67, 73, 75, 111, 118-120
Prostaglandins 59, 67, 101, 106-108
Protein kinase C (PKC) 6-11, 25, 36-38, 40, 54, 61, 62, 64, 65, 98, 100-106, 108-112, 121, 125
PTEN 85, 89-91

R

Rab 26, 72, 78, 79, 98, 125, 133, 135, 140, 141
Rab5 39, 78, 135, 139, 140, 143
Rap1 72, 76-79
Respiratory burst 2, 58-62, 67, 98, 100-102, 107-111, 122
Retinal epithelial cell 4
Rho 2, 10, 12-17, 25, 36, 54, 72-77, 79, 87, 103
Ryanodine receptor (RYR) 121

S

Salmonella typhimurium 73, 75
Scavenger receptor 4, 7, 9, 11, 23, 97, 119
Sec61 26
Serine/threonine kinase 1, 2, 10, 11, 17, 100, 112, 125
SH2 24, 27, 34, 35, 86, 88-91, 100, 101, 111, 112
SH2 domain-containing inositol 5' phosphatase (SHIP) 27, 40, 85, 87-91
Shigella flexneri 75
SHP-1 27, 40, 85, 87-89, 91, 92
SHP-2 27, 40
SIRPa 27
Soluble NSF attachment receptor (SNARE) 39, 125, 133-135, 138, 139, 144
Sphingomyelinase 99, 102, 108-111
Src family 6, 7, 9, 15, 16, 24, 34, 55, 101, 102, 105, 124
Src homology 2 (SH2) domain 40, 86, 112
Syk tyrosine kinase 35

T

Talin 137
TGF-β 27, 28, 59, 60
Tissue remodeling 1-3, 16, 27, 58, 59, 66
TNF-α 54, 87, 98, 109, 110
TRP superfamily 121
Tyrosine kinase 1, 2, 4, 6-9, 17, 24, 25, 27,
 28, 34, 35, 73, 75, 86, 89, 100, 111, 120,
 122, 124

V

Vav 9, 13, 36, 37, 73, 87, 89
Vinculin 137

W

Wiskott-Aldrich syndrome protein (WASP)
 13, 15, 16, 36, 74, 76, 136
Wound healing 2, 3, 16, 58, 59, 66

Y

Y. pseudotuberculosis 75
Yersinia enterocolitica 75